MOLECULAR
BIOLOGY
INTELLIGENCE
UNIT 16

# The Genomic Potential Hypothesis:
## A Chemist's View of the Origins, Evolution and Unfolding of Life

Christian Schwabe
Department of Biochemistry
Medical University of South Carolina
Charleston, South Carolina, U.S.A.

CRC Press
Taylor & Francis Group
Boca Raton London New York

CRC Press is an imprint of the
Taylor & Francis Group, an **informa** business

# The Genomic Potential Hypothesis: A Chemist's View of the Origins, Evolution and Unfolding of Life

**Molecular Biology Intelligence Unit**

Designed by Jesse Kelly-Landes

CRC Press
Taylor & Francis Group
6000 Broken Sound Parkway NW, Suite 300
Boca Raton, FL 33487-2742

First issued in hardback 2019



ISBN 13: 978-1-58706-044-1 (hbk)



**Library of Congress Cataloging-in-Publication Data**

Schwabe, Christian
The genomic potential hypothesis: a chemist's view of the origins, evolution and unfolding of life / Christian Schwabe.
p.cm. -- (Molecular biology intelligence unit; 16)
Includes bibliographical references
ISBN 1-58706-044-2 (alk. paper)
1. Life -- Origin. I. Title. II. Series.
QH325 .S386 2001

# CONTENTS

# ACKNOWLEDGMENTS

It is done and I am grateful to all who spent untold hours to make it come alive.

Jutta Schwabe has read the manuscript many times (comma, period, paragraph, and all) and has, with untiring critical facility, pointed out what words it would take to stay in contact with my potential readers. My colleague, Erika Büllesbach has given invaluable computer support, particularly as concerns illustrations, and was instrumental in the production of critically important proteins. Marion Hinson provided expert secretarial support expeditiously and on a moment's notice. Ron Landes has kept the presses running through several cycles of revisions and his confidence and patience have been an inspiration to move ahead, to finish and to face the world.

# PREFACE

Why a new paradigm of evolution just when everything seems to be well under control? The courts have come down on the side of the current evolutionary paradigm, and Monday through Friday all youngsters will hear the unadulterated story of the slow transmutation of lowly creatures to something like their parents. Ideas are sort of "in the air". In the mid-nineteenth century several chemists recognized a certain continuity in the properties of the elements and began to organize them in a systematic way. Dimitri Ivanovich Mendelyeev (1834-1907) saw a natural law reflected in this order and made predictions concerning new elements which were actually confirmed during his lifetime. The discovery of calculus by both Isaac Newton (1642-1727) and Gottfried Wilhelm von Leibnitz (1646-1716), and the simultaneous formulation of the concept of natural selection by Alfred Russell Wallace (1823-1913) and Charles Darwin (1809-1882), are some of the many well-known examples recorded in the history of science. Today there are those rumblings again that promise change.

The years between Lamarck and the first printing of Darwin's *Origins of Species* in 1859 were awash with evolutionary thoughts. Etienne Geoffroy Saint Hilaire's unity plan of animals and his brand of higher philosophical anatomy was exported from Paris to London where a leading British comparative anatomist, Robert Grant, formulated ideas about development and metamorphosis and actually talked about evolution in the Darwinian sense years before Darwin published his theory. The research of Adrian Desmond (*Politics of Evolution*, University of Chicago Press) again reveals fragmented awareness put together into a coherent picture, not by evidence but by the powerful imagination of one person, Charles Darwin.

Evidence was actually against Darwin as it was against the pre-Darwinian school of transmutationalists. In the highly politicized climate of London in the 1830's the ideas of Jean-Baptiste Lamarck (1744-1829) replaced divine forces (and privileges) with an egalitarian system whereby the environment essentially determined the destiny of individuals. Suddenly evolution had become the backdrop for a fight against oppression by the privileged class. The highly regarded British anatomist Richard Owen, who was part of the establishment, argued counter to the new movement that the fossil record is static and that no evidence exists which would support the transformation of one animal into another. A quote from Desmond may be appropriate at this point.

*He (Owen) showed that Geoffroy's anatomical sequences ran contrary to nature. "Ichthyosaurus, Plesiosaurus, and Teleosaurus are genera which appeared contemporaneously, one neither preceded nor came after the other" he said. "Moreover the Ichthyosaurus could be traced generation after generation without any sign of change and they disappeared in the chalk as they suddenly had appeared in the lias." There was no succession. Only if animals were pulled out of their chronological order could a sequence be artificially constructed.*

It may be argued that Owen leaned toward reasserting God's continual creative power, but the transmutation post-Lamarckian movement was equally beset by ulterior motives of a political nature. On the other hand, Owen's observations (dinosaurs were discovered and named by him) were real and can still be confirmed today. The fossil record must be recognized as reality and so it is in this book. For explanations, however, chemistry and chemical determinism will be called upon to replace Darwin's emulsion of luck and biology and Owens divine intervention.

Darwinism held up for 150 years, but evidence and conceptual advances have caught up with it and a struggling paradigm tends to intensify a search for a new one. Darwinism, including its 1930 update (neo-Darwinism), was built upon a wrong premise just like Newtonian astronomy was, and all the reverence we have for great minds does not count toward the explanation of nature. Our planets are not captives of an attractive force and evolution is not driven by natural selection. Newtonian attraction is mimicked by space curvature, and evolutionary trees are an illusion created by the chemical principles that guided the construction of primordial genomic material. Evolution is a valid concept as it relates to the appearance of complexity but the process cannot be driven by Darwinian success parameters. This philosophical disparity is serious; it will haunt us until we react.

According to a report in *Scientific American* by Rebecca Zachs (Oct. 97) science teachers are upset about their inability to deal with objections concerning the constructive role of mutations other than to label them misinformation. The students are correct, the paradigm is definitively wrong on this point. Would it not be embarrassing to have to realize that Gallilean minds might be stuffed with Ptolomean concepts? We simply must try to find a viable model that will lead us through this century.

Ernst Mayr, the foremost living interpreter of Darwinian philosophy, is removing the evolutionary idea from the fundamental laws of science and declares it "concept driven". He is distancing himself, Darwin and his evolutionary theory from Laplacien determinism and thus makes Darwinism untouchable by Popper's falsification test for hypotheses of science. Darwinism must be recognized as a scheme of plausible explanations, each justified by a prior assertion. Ernst Mayr is correct as concerns the character of the Darwinian model and with that realization the answer has been found as to why a new hypothesis of evolution.

Laplace's ideas, however, have never been defeated and cannot be denied because determinism is merely a restatement of the laws of cause and effect which is valid for the nonreligious, nonpolitical world. (What in fact happened is that determinism has been 'voted out of office' by political groups). In concept all deterministic processes are predictable, in practice some are too complex for our brain power to fully comprehend.

The Genomic Potential Hypothesis develops the evolutionary events as a consequence of chemical/biochemical principles. Even complex biochemical processes such as thought and emotion fall under the purview of science and perusal of contemporary science literature supports that statement.

Why a new hypothesis? How many answers are there for every question? H.T. Buckle (1821-62) suggests that, "There is a spiritual, a poetic, and for ought we know a spontaneous and uncaused element in the human mind, which ever and anon, suddenly and without warning, gives us a glimpse and a forecast of the future, and urges us to seize the truth as it were by anticipation...", and it may be just as good to leave it there.

# Introduction

The Genomic Potential Hypothesis is a biochemists' view of the origin, evolution, and development of life. Large numbers are second nature to a biochemist and though he rarely ever thinks of it explicitly, the concept of mass action is a part of the definition of chemistry. The origin of life, from that perspective, will turn into an event that occurs on the molar scale in units of $10^{23}$ and is driven not by needs of biological systems but by mass action, energetics, structure, and kinetics. The outcome of every reaction will be a normal distribution of compounds around a major group and progression in self-assembly will be a shift from one major group to another. Nucleic acids are a bottleneck, a gate that herds the mixtures of chemicals onto the path to life. Beyond those requirements chemistry will go into every favorable direction, and complex functions that are uniform in all life forms and which have been considered frozen accidents in the past are, to a biochemist, testimony to the fact that no alternative assembly was possible.

This approach to evolution entails the total denial of constructive accidents. Mutations are a reality and while most of them are of no consequence or detrimental, one cannot deny that on occasion a beneficial mutation might occur. However, to invoke strings of beneficial mutations that suffice to reshape one animal into the shape of another is not merely unreasonable, it is not science. As one might expect, a major postulate based upon the new hypothesis is that species are the products of normal distributions of nucleic acid sequences that were produced separately in the primordial biogenic foci, they are the chemical heritage of each species. Evolution will be restricted to the reorganization of nuclear material in line with equilibrium constants and kinetic parameters that govern the quasi two-dimensional chemistry of nucleic acids. The actual evolution occurs at the cellular level and is noted only by the results appearing in the fossil record as small versions of the final form.

Nucleic acid is the conductor as well as the memory for the living cell, which is a remarkable fact with deep cutting consequences for the limits of any hypothesis of evolution! The memory has built itself without anything to remember, no goal, no survival drama, just following the mass action law.

The arguments given in this book question the old explanation in order to make room for new thoughts at the sight of the same evidence. It is widely accepted that there is no way to proof a hypothesis, but a current hypothesis can be disproved when science has driven development beyond the foundation of the old model. So it happens that the same data will be presented with a new interpretation and that too is not uncommon in a world that was mostly flat not too long ago.

Heavy reliance upon the abiotic or immediate prebiotic period makes it necessary to discuss many aspects of chemistry that may be, in a different context, quite familiar to many of my readers. In the new paradigm all of the relatedness arguments must go all the way back to that one and only biogenic period that occurred for a limited time in the post-accretion world which was a period purely of chemistry. The early earth was not just tolerant, it was the patron of chemical self-organization of life.

A few general prejudices must be addressed. In fact, some ancient ideas might return in a new form, which might cause some heckling. So I must ask, would it fill the reader with suspicion if the fossil record instead of being a burden, with the missing intermediates and all, would become prime evidence in favor of the new concept of evolution? During Newton's time mercury's orbit was a problem for scientists but should it make us suspicious that in modern astronomy it has become crucial evidence for Einstein's view? Modern atomic theory, is it suspect because the Greek philosopher Epicurus already enunciated the concept of atoms 2000 years ago? Now, is the genomic potential idea suspect because 2000 years ago Christians for unrelated reasons insisted on special and separate creation which superficially resembles the themes and variations of chemistry? Why not simply admit that our ancestors guessed right. The modern approach to natural phenomena is based upon completely different reasoning in every case. The difference is that modern science is pulsing through the tissues of the newly-grown intellectual constructs. Here is the new evolution that I would like to share with you, brief as it is; the book will provide the explanations.

In the Genomic Potential Hypothesis every species had its own origin and its own world-line that connects the chemical origin with an extant or extinct life form. Billions of small puddles of fresh water contained normal distributions of nucleic acid polymers that formed themes and variations of potential coding material before the oceans were established. Oceans were not biogenic, they were infected with life by the small foci of biogenic chemistry. These foci were the origins of species and their variants. Chemistry provided the syntax and one set of letters, biology took 3 billion years to write a plethora of stories.

Evolution is the ripening of the embryonic quasi stem cells of each origin which began to transform, group by group, into the final phenotype in the Cambrian, the least complex ones being first to make their fossil imprint. Once established, species do not branch or adapt beyond physiological limits. Stressed beyond these limits a species will suffer extinction. Mutations are not a mechanism to produce new organisms. Therefore there are no intermediate forms and the evolutionary trees are an image created by the sequential ripening of pro-forms and their rapid rise into the fossil scene.

Life is a fantastic story, but what precedes it is every bit as incredible to the human mind. The innate properties of chemistry are laid down in the first three minutes of our universe, and a hole, torn into the fabric of space by the pristine explosion, gives us time and a gravitational frame and awareness to know how fragile and fleeting we exist under the temporary protectorate of a small and unusual planet. Everything derives from universal concepts, it would create a discontiguity were one to exempt the hypothesis of evolution.

# Chapter 1

# Life in a Tenuous Universe

The question is, do we live independent of the shape of our universe, or are we an intimate partner as well as a beneficiary of its peculiar structure and its awesome dimension? The space-time that life needs to exist has been torn out of matter energy in the core of a vast and expanding geode. We are creatures as well as prisoners of space and the vagaries of time. Our home, what is it like, are we (living creatures) a quirk or a consequence?

Could there really have been nothing? It is hard to imagine, yet the standard model of the Big Bang starts with a submicroscopic space of very high density. That does not mean that it is true, but it is the model that arises from doodling with Einstein's equations and Einstein is the best bet around.[1]

Relativity tells us that time slows down in extreme gravitational fields (around a black hole, for example) and during extreme acceleration (this is referred to as time dilation). The primordial singularity is imagined as being much denser than even a black hole and if time at the Schwartzshield radius (the sphere of "no return" around the black hole from which even light cannot escape) must stand still, then time must certainly have been captive of the gravitational force in the Big Bang singularity (singularity here refers to the unity of all forces as opposed to the uniqueness of our universe). The perception of time, it is theorized, would be the same as ours for anybody living (hypothetically) in a black hole. The light would move away from the hole at the speed of light measured in unit time in the black hole! But time does not flow at all, or infinitely slowly under these conditions and hence—for the outside observer—light does not get out. This is part of the invariance concept of relativity. Perhaps it is better to say that time is not defined during this period. Physicists prefer it that way! When they face the stark consequences of their own ideas they fade from the scene much like (Goethe's) Dr. Faust did at the sight of the spirit of the earth whom he had conjured.

Time is recognized by progressive events and space is recognized by distances. Purportedly the early universe was much smaller than the size of an atomic nucleus, minuscule compared to its size of 20 billion light years which is often called the visible universe (based upon the assumption that there is an invisible one). This is another interesting subject for intellectual "doodling". We have no idea what physical form a sub-atomic-size universe would have assumed, but according to physicists it could have been pure unpartitioned energy. Physicists are searching for a unified field theory dependent upon the (probably correct) assumption that just before the "Ur Explosion" all forces were one and the same. The creation of space segregated forces and created time. Dr. Fred Hoyle and Dr. Arno Penzias once explained to me that I must not imagine the Big Bang singularity as something one could stand next to and photograph (conversation during the Welch Foundation Symposium on cosmo chemistry, 1981). There is no space next to the Big Bang singularity, all space is within it and all time as well. In retrospect I think that this is the physicist's way of discouraging further questions.

There are black holes in the universe and we can live very well next to them, albeit at a respectable distance. If we could put a clock into the black hole and if we could find a way whereby this clock could remain intact and signal time progression to us we would notice that, compared to our clock, the black hole clock stands still. The same should have been true for a clock within the primordial singularity except there was purportedly no outside time to which to comparä%oit. In any event, we can imagine that the Big Bang explosion was 'planned' for twelve o'clock and that it is one billionth of a second before twelve; will this explosion occur? This question is justified since even atoms in the atomic clock "feel" the slowdown. Radioactive decay should stop in extreme gravitational fields. An explosion without molecular motion is unthinkable. The particles within the primordial singularity, or a black hole for that matter, feel like the clock in this super-gravity field. How can anything happen under such conditions? Only if time, like quantum phenomena, can tunnel through energy-barriers is it conceivable that the Big Bang might have occurred (as we fairly well know it did). The initial phase of the explosion has been described by Stephen Weinberg[2] as a three and one-half minute phase and, as there was no space or time outside that explosive event, we must presume he means the time as it would be measured at the local scene. On our clocks the time might have counted 3 billion years, and the explosion should really have seemed to occur at the speed of an extreme time-lapse movie if the relativistic time dilation is real to its final consequence. Indeed, if such an explosion occurs time must instantaneously start running (but how fast?).

By our reckoning a light beam travels 300,000 km/sec. Yet an observer sitting on the leading edge of a beam would report that light travels through the universe at an instant because time does not progress for the light beam.

What is the reality in each case? We can measure the arrival of a light beam on a target so this perception is correct —for us. Is time a dimension of motion primarily and of space secondarily? Nonsense, of course. There can be no motion without space and no space without time.

Thus, we must also argue now that the extremely small, though finite, space of the primordial singularity could not have existed without a finite, however little, time. Having a million dollars at no time means one is not a millionaire, and a singularity at no time is no singularity. What does it mean if one says time stands still? It would be very difficult to imagine, perhaps about as difficult as it is to picture a storm at rest. What is time at rest? To make air masses of a calm into a storm we need a pressure gradient, what could make time out of its progenitor? A mass energy gradient that produces gravitational force or acceleration which is its physical equivalent? A pristine explosion?

It is said or implied that the origin of this pristine explosion (pristinus, the earliest, untouched, uncorrupted), ordinarily referred to as the Big Bang, was even denser than a singularity in a black hole, but I have never seen any theory that would explain anything more dense than the point where all space, shape, memory, and all forces that we know have become one and the same. And if space and time are buried within the singularity then the uncertainty principle should also no longer prevail. But if the uncertainty principle does not prevail then quantum physics must also be invalid and the singularity could never advance to a point where it should explode. We have again reached the point where we must say that it did, at least according to observations in the present universe, such as the background radiation. The situation is somewhat reminiscent of biochemical evolution which, by and large and in reference to our time-frame of experiences, is completely impossible, and yet we have to find a path to explain it because it stares out of and into our faces. At any level of lack of conviction let us give in then, and say that quantum phenomena were still occurring at the singularity level and that occasionally a large singularity, a large black hole, could explode in spite of the fact that Stephen Hawking has calculated a fair chance for an explosion to exist only for a minuscule black hole, and even those have not been detected yet. Could the pristine explosion have been due to a black hole in an existing universe, and in contrast to what Penzias and Hoyle said, you could stand next to it and photograph it if you had the fortitude or misfortune to live in the old, cold and dying universe and have a telescope trained to the right spot in space? Was the recent incredible outburst of energy from a galaxy about 15 billion light-years away the birth of a new universe? A sister universe because the explosion we are registering now would have occurred when we were 15 billion light-years closer together. The black hole would thus explode within an existing universe to form another bubble, and it is not totally impossible, at least in our imagination, to resurrect (up to a certain point and in a different way) the steady state universe[3] except that new matter would not come from the center but peel off the inner surface of the expanding edge like amethyst crystals in a geode. In such a picture the big crunch would be replaced by the slow accretion of material into the black holes that are apparently at the centers of large galaxies which could explode again, perhaps at the time when the previous universe had reached a temperature of maybe $1/10^{50}$ degree K (just to pick a number). It would be a geodesic universe, wherein geodes would be produced within geodes on a fantastic time scale and where every universe would be unique, singular, and forever unaware of any other universe. The advancing edge, traveling at the speed of light, is tearing apart space within the bubble to the point where time can become its meaningful adjunct. If there is space, i.e., a point to the left and a point to the right, then there has to be time to travel between the points. Light becomes not only the prime messenger of space but is created by space and by the lower gravity which is a consequence of exploded space. In the first million years of this universe light could apparently not travel. It would have been reabsorbed by the thick particle soup according to the "standard model". It must have been an eerily dark explosion.

Here we are then enjoying the temporary expansion of space and time in which all our chemistry can occur and in which evolution has occurred by the laws of chemistry. The space between galaxies is not empty (virtual particles can appear in a vacuum), and it appears not to be unstructured, and the physicists are speaking about a wormhole configuration whereas anyone with a sense of esthetics would have wished they had thought of Swiss cheese to go along with quarks (Farmer's cheese in German) which appear to be the fundamental particles of matter. But no such luck, our life has developed in wormholes if they will be confirmed. Terms like this have a tendency to stick. Before wormholes the space between galaxies was an ether, a beautiful term whose time may never come.

Are we confined to a bubble like the quarks are confined to the protons or neutrons? Is this a gravitational confinement? Physicists say no! Gravity plays no role within the nucleus. Strong and electro-weak forces play a role; gravity is extremely weak in the short range. Atoms feel gravity—but they are held together by other forces. Quarks can move without constraints inside their little bubble but, as soon as they move toward the periphery, an ever increasing force is restraining them[4] Does the proton have an edge?

We can move in our universe, our bubble of expanded space time at comparatively slow velocities but, as soon as we would speed up to anything only nearly the speed at which the edge of our new universe expands into an old one, we would feel the same restraint. The physicists do not like the concept of an edge but they have nothing convincing to put in its place. The energy to move a quark out of a hadron would be so great that a string of new quarks would be produced. The energy to accelerate us up to something near the speed of light would cause a prohibitive increase in our energy. We are velocity-restrained. No matter how fast we are moving (in terms of attainable speed) our universe becomes larger at all times and the high-speed traveler will recede constantly from the edge he is trying to reach. How about if we then moved in the opposite direction, away from the nearest edge of our own expanding bubble? The same thing would prevail because now we are beginning to match the receding edge on the other side of our bubble which would be just as impossible.

In such a universe the expansion function would be precisely 1, i.e., no crunch or time reversal would occur except in black holes,[5] our time sink. Near the edge of such a universe there would be the great wall of galaxies uniformly spread like the background radiation. Could we fly faster than the speed of light (the quasi escape velocity from our universe), would we then penetrate the great wall and the receding edge and end up in the previous universe?

I can see the physicists' smirk at these attempts of a chemist to find a proper place for our evolutionary history. Consider this my revenge for the physicists' excursion into biological evolution. It is not a full revenge, for at least I am trying to transmit what an unadulterated mind can make of the stories they are telling each other. The revenge is incomplete because it is to me a most disturbing experience when I see world-class physicists in their popular writings mindlessly repeating our impossible paradigms of evolution, collectively known as the New Synthesis. Physicists are not particularly timid. When Stephen Hawking[6] had finished presenting his paper on some aspect of the thermodynamics of evaporating black holes, one fellow physicist got up and said something similar to: "Interesting, but you know it is all rubbish" (which it was not).

When a Darwinist tells them that the living world began with one cell put together by luck and against the odds of $\sim 1{:}10^{348}$, and that all other organisms are derived from that cell by a string of millions of lucky mutations, each against the odds of $1{:}10^{300}$ (just to keep the numbers small), they look on, starry-eyed, as if the biological world were something out of this universe. Confronted as to the validity of such concepts they step uneasy from one foot to the other mumbling paternalistically something about quantum phenomena and uncertainty instead of calling it what they would call it if they were to argue with a physicist—rubbish (which it is)!

Fortunately we shall live on regardless of which idea describes our universe and our story of evolution. We survived Joshua's flat world, Ptolomy's circles, Newton's attractive force, Lamarck and Darwin. Ideas do not change the cosmos, they only change the way we look at it. Yet the challenge remains to find what guides all of it, space, time, forces, chemistry, and life.

Look at the big not-yet bang during the 'million year second' just before it goes off; somewhere there, in a miniscule spot within nothing, is the milky way, your sun, the earth, your home town and you, your cat, all of it is there already, for nothing enters the universe and nothing leaves. A cosy thought isn't it, yet, something is not quite right. Is it, I wonder, just possible that this is the mathematical endpoint and that reality parts ways with mathematics before we are quite there? What bothers me much is that the 'standard model' needs to be modified by a 'faster than light' inflationary period[7] for which there is no mechanism, no foundation, no reason other than that it helps make the standard model match observations. The need for this epistemological *faux pas* indicates to an observer that the universe may in fact have started a little larger than the model demands. There is some consolation in the fact that our theories do not change our living space.

The universe is a fantastic place by any standard and, most incredibly, we are a product of its formative forces. Whatever made time and space made us! Biology needs space, time expansion, and light, which are at once the causes and results of cosmogenesis. Life will die with the old universe, it will rise again with the new one, and it will again be a short episode of self-consciousness for a generally inanimate marvel.

## References

1. Einstein A, Infeld L. The Evolution of Physics. New York: Simon & Schuster Inc., 1938.
2. Weinberg S. The First Three Minutes. New York: Basic Books Inc.,1977.
3. Bondi H. Cosmology. Cambridge: Cambridge University Press, 1961.
4. Riordan R. The discovery of quarks. Science 1992; 256:1287.
5. Rees MJ. Black holes in galactic centers. Sci Am 1990:56.
6. Hawking SW. The edge of spacetime. Am Sci 1984; 72:355.
7. Guth AH. Inflationary universe: A possible solution to the horizon and flatnes problems. Phys Rev 1981; D23:347.

# CHAPTER 2

# The Frame for New Hypotheses of Evolution

How does one present any new idea which is in principle impossible to proof? Hypotheses are self-limiting and an old paradigm of evolution will fall victim to its errors in logic. In biology in particular (because it is so close to our skin), old paradigms tend to exchange fundamental guidelines of science for political correctness and technological expedience. Not for scientific restrictions, but for political reasons it would be difficult to find an unbiased jury for the proposal that humans might be of various origins and, in part because of methodological convenience, the computer-oriented molecular biologists would (and do) enthusiastically ignore the evidence that protein structures may not bear a parametric relationship to genealogy. Science, culture, and vested interests tend to fuse as paradigms age and, off and on, the science must be extracted from this matrix and held to the light to see whether it holds up on its own. To that extent one must have clear ideas about what can be known, what constitutes evidence, what are truly dependent variables, as opposed to wishful thinking induced by emotional needs. Ah, but mathematical models will help, unemotional mathematics will brush out the rubbish of prejudices and human sentimentality, or will it?

Mathematical models that describe and predict the inanimate world quite well are actually of little value in the system of deterministic chaos that governs biology. The answers one can expect from mathematical approaches to evolution (in contrast to my earlier perception) cannot be narrowed to less than the surface of the chaotic attractor of the system which is a little like watching evolution on earth from a satellite.[1] The limits of the attractor surface are given by the initial conditions which are not knowable in sufficient detail.[2] Empiricism can help, after all our laws of science by and large are the results of repeated observations.

Fossils and chemistry would be stretching correlations within a Darwinian framework but that relation is the substance of the new hypothesis. Chemistry is part of the logical core of biology, which will become more obvious when speciation is considered in the Genomic Potential Hypothesis. It is, like all of evolution, about molecules that self-organize without the help of anything but other molecules, i.e., chemistry.

Chemistry is invisible until one can make a prediction based upon what chemistry can do and then look for the results, which in this case are clearly visible in 3.5 billion year-old rocks. There was nothing but chemistry before so that the time and location of imprints of early microbes near the edge of existence of our planet takes on an unusually clear meaning, i.e., mass action laws of chemistry left them there. For that most forceful reason a hypothesis about biogenesis must include an account of a large number of origins at time zero which designates the few million years of biogenesis.

A cell is extremely complex by any standard, but as a unit in the path of development of large forms of life it again submits to a simple logic. For example, anyone starting with a single cell at the origin of life will end up with a Darwinian/Lamarckian model of evolution. There is no escape because a singular start must branch incessantly to give rise to the manifold forms of life all around us.[3] Multiple origins instead would lead to a polyphyletic, a polyclonal, model such as the Genomic Potential Hypothesis.[4] Bacterial fossils are found in ancient rocks from all corners of the world,[5] and, although the search has barely begun, for the thoughtful observer the results point to models that predict multiple origins of life. The Hadean edge of life is a rather convincing, if unusual, demonstration of the mass-action concept of chemistry.

The stability of microbes presents another set of conceptual limitations for hypotheses of evolution. Here stability must be understood not as stasis but rather as the ability to reproduce accurately $10^{14}$ times approximately since the beginning of life. Such fidelity in phenotype reproduction is incompatible with models that call for organisms to acquire all novelties by random mutations.

Chance is a very popular, if often misunderstood notion, which does not exist in the world of causality. In terms of human perception chance processes are events that have an unknowable history and are not predictable. The chance processes often referred to in scientific discourse describe the diffusion of reactants through solvents, which is subject to Brownian motion and therefore a function of temperature and concentration. It follows that in concentrated solutions the diffusion part, the chance portion, cancels out of these equations because collisions are always much faster than the overall reaction. Thus, in close quarters reactivity becomes a function of the structure, the bonding angles, bond strengths and other thermodynamic properties of the molecules, and in that context the reader will remember the fact that molecules will react as directed by structures. Even if multiple components are present in one container, they will still react according to thermodynamic and kinetic paths for each constituent and the structurally most favorable reaction will be dominant with proportionately less of the others produced. This, the best-documented behavior of molecules, carries in all its simplicity the recipe for self-assembly and explains why the oceans were not the place where life originated. Atomic structure and high concentrations are nonnegotiable conditions for the conceptual path to the biosphere.

The aggregation of complex systems always proceeds through simple steps. Simple processes form complex systems by going from one stable plateau to another before the summit is reached. Nucleic acids are the first plateau for the ascent of life. A second stage is the genetic code, the virtual frames that subdivide the endless gene into segments of three, each to signal the addition of one of 20 amino acids at a time to a protein chain.[6,7] The structure of the genetic code and its universality throughout all of life was a great discovery of the 20th century, but what is the science of it? The opinions of Darwinian biologists run much in favor of the single origin explanation (a mechanical one) whereas a genomist chemist observes that such an extraordinary uniformity means that the reactions leading to the genetic code must have been highly specific. This crucial difference must be incorporated into a new model of evolution.

Evolutionists by and large acknowledge the role of chemistry in the origins of life but fail to realize that the multiplicity of origins is a corollary forced upon us in turn. As a consequence of this corollary, variety becomes an initial condition and not the result of biological meandering; it is a function of the breadth of primordial chemistry, which enters the equation of life as a nonrenewable resource. This endowment, theorizers must remember, diminishes constantly during biological processes, and finally ends in extinction of species after species with or without asteroids. The prominence of extinction, some times illustrated by claims of 99.9% species loss since the Ediacara,[8] strongly endorses the idea that the production of variety may not be an attribute of biology. It appears that biological systems cannot produce targeted mutations to overcome environmental restrictions yet, that concept is popular in spite of the fact that species will unceremoniously disappear when conditions vary beyond the competence of its variants. Chemistry does not feel direction if the need is not a thermodynamic one, which means that there can be no hypothesis built upon a "need response" of organisms. Furthermore, integrated systems do not tolerate significant changes of core functions, and that is why mutations kill if they do anything.

The chemistry of the origins of living systems is a complex problem which emotionally invites the thought of uniqueness and, when the current model of evolution gave in to the desire for a singularity, a host of well deserved problems began to drag it down. From the unique Darwinian origin, diversity has to arise by reproduction with variations. A hypothesis built upon the 'unique origin' idea has to produce intermediates in the fossil record as well as in terms of protein structures. This was well recognized and led to the neo-Darwinian ideas of molecular genealogy which derives plausibility from the fact that in some way the phenotype must be an expression of the genotype, and from the assumption that gene duplications and mutations are the basis of species development.[9]

An absolute condition for molecular genealogy is a single origin of life and a uniform rate of mutation accumulation. These conditions are not confirmed by fossils or molecules and therefore the idea of lateral gene transfer has been recruited to keep the hypothesis afloat.[10] Proposals of uneven rates of molecular changes as a function of time do not even raise eyebrows any longer. If indeed the single origin concept were correct, then lateral gene transfer would make it unfalsifiable because it would totally obscure the origin of a protein and thus remove from the hypothesis this prime criterion of science! That is the kind of help no hypothesis can endure. It is a case where solving a local condition causes global problems which lead into the booby traps of ad hoc hypothesizing. Every time a protein is found where it should not be, lateral gene transfer will be invoked. Transfer between out-of-sight partners, such as pigs and tunicates have been suggested in order to protect a hypothesis and that, disregarding the comical aspects, is against progress in theoretical work.

The statement of principle of any hypothesis excludes mutations as a constructive evolutionary mediator. Mutations are causal in terms of the chemical events, but they are not contiguous because the history of an organism has no influence on them. It is the lack of contiguity that banishes mutations from a constructive role in evolutionary science and no amount of trimming will make this hoof fit Cinderella's shoe. There is no deterministic component in the modification of a genome by a nonspecific agent that relates to the life cycle of the organism. Random mutations may be tolerated by micro-organisms, but they are too destructive for multicellular life. While mutations as (often fatal) mistakes are well known among humans,[11] as active, constructive components they are as unknown in practice as they are inadmissible for a hypothesis of science.

The multiple origins hypothesis is built upon the marvelous but not miraculous properties of chemistry which are engraved in the structure of atoms. All of the postulates of the Genomic Potential Hypothesis, which stay well within the epistemological framework of a scientific dissertation, provide a legitimate basis for experimental testing.

## References

1. Mandelbrot B. The Fractal Geometry of Nature. New York: Freeman, 1976.
2. Gleick J. Chaos. New York: Viking Press, 1987.
3. Darwin C. The Origin of Species by Means of Natural selection. New York: The Modern Library of New York, 1859.
4. Schwabe C. Evolution and chaos. Computers Math Applic 1990; 20:287.
5. Schopf JW. Science 1993;260:640.
6. Nierenberg MW, Matthei JH. Proc Nat Acad Sci USA 1961; 47:1588.
7. Nierenberg MW, Leder P. Science 1964; 145:1399.
8. Raup DM. Biological extinction in earth history. Science 1986; 231:1528.
9. Shapiro JA. Adaptive mutation: Who's really in the garden? Science 1995; 268:373.
10. Fagan T, Hastings JW, Morse D. The phylogeny of glyceraldehyde-3-phosphate deydrogenase indicates lateral gene transfer from cryptomonads to dinoflagellates. J Mol Evol 1998; 47:633.
11. Erdjument H, Lane DA, Panico M et al. Single amino acid substitutions in the reactive site of antithrombin leading to thrombosis. J Biol Chem 1988; 263:5589.

# Chapter 3

# Genomism and the Nature Trail

As genes move into the center of a hypothesis one needs an "ism" to refer to the background of ideas that make up the new model. Thus, as the term Darwinism describes the single origin, the descent with variation, adaptation and constructive mutations and so on, genomism describes a system centered upon large amounts of abiotically synthesized genomic raw material which limits not only the developmental potential of each species but also the level of genomic complexity on a planet. How to get evidence for the new model is the question.

Genomism had a fairly unassuming start close to 30 years ago during a small group session on the topic of Darwinian evolution with refreshingly bright students. There was a moment at the blackboard when it seemed that all the incongruities of the old evolutionary story that had been plaguing me off and on came into focus and that made me hesitate for a moment until an explanation that had been sort of on the backburner suddenly became clear. It was a tense moment but my students thought of it as a setup, a teaching technique. I remember writing ACTG, which stands for the nucleotide bases in several permutations, and starting a discussion as to how frequently any sequence we might want to write could appear if we had a mole or 1000 moles ($\sim 6 \times 10^{26}$ bases) available to select from. The discussion drifted for a wonderful hour from gene duplication to redundancy which would make that duplication unnecessary, eliminate mutations and make altogether for a happier way to produce variety. Redundancy and reiteration had conned scientists into drawing evolutionary trees from protein structures and that, in turn, provided the impetus for matching these trees to those that the Darwinians had designed for species identification. If, instead of adapting molecules to the purported speciation scheme one were to adapt the sequential appearance of animals to the chemical production of redundancy, one would eliminate the need for unique events and for the intermediate forms that the fossil record will not yield, in spite of long and intense searches.

Where to go from here, how to obtain evidence for these ideas and how can one know that the blackboard discussion it is not the harebrained scheme I was assured it was after the first contact with members of the old guard. Clearly, there is no experiment that can test the grand picture of evolution and the evidence is strewn everywhere without labels. Now, would it be time to join the searchers that are tilling the landscape for evidence?

Away from all this, but not too far, is a fossil bone fragment, buried in a two-million-year-old earth formation, which may have belonged to one of our purported 'ancestors', *Homo habilis*. Someone else contends that this was not *H. habilis* but rather *H. erectus*, pointing at the facial angle and the teeth. The discoverer again points at the facial angle and the teeth and declares it to be *H. habilis*. Both contestants had in their minds a picture a likeness of a hominid, plus an idea where it came from and to what creature it would be an ancestor. The process is well described by Johanson and Edey.[1] Finally they decide to let the bone speak for itself, but that produced absolute silence. That was a noble gesture without merit because only through our imagination can objects become evidence.

The crux of the matter is that we have to make the evidence talk through us and that therefore hypotheses are born more of introspection than inspection; the Genomic Potential Hypothesis is no exception. Darwinism was finished before Darwin embarked upon the Beagle trip that brought him the kind of personal suffering that folklore associates with vision. In fact Darwin's imagination was not in harmony with what he observed, but the idea was stronger and science was not far enough developed to provide an alternative that could have prevented a fall-back to divinity. Today, I think, Darwin would have enjoyed a tour through air-conditioned libraries over a steady ground, and he would be the first one to concede on the basis of evidence collected by others. Henry Huxley, of course, would have been invited as well, and on this quiet walk over thick carpets more than once one would have heard him mumbling in the back, "I told you so Charles." "But all had different beaks, Henry." "Indubitable, but all were still finches". In the long run, the mounting evidence (here meaning the logic of progressions rather than bones) will edge out the old view, but as things are standing right now Huxley's talents would be invaluable.

If evidence has so little authority about it because it cannot come to life but through our imagination, then one must admit that ideas are the core of our drive toward discovery. Theories cannot be proven and they remain true until falsified by factual and conceptual incongruity.[2] Evidence has a crucial role in denying a hypothesis, but if observations are not so reliable on the "pro side" how could they be more authoritative on the "nay side"? The proposal that X came into existence via path Y requires one to exclude all other possibilities. Conversely, the statement that X is not produced by path Y requires one merely to exclude Y, which is in principle much less involved.

The new hypothesis is compared to the evidence that has been assembled by the efforts of paleontologists, biochemists, and biologists, and was deposited in our libraries. Hypotheses, however, are not summaries of findings but are syntheses based upon ideas that must be tested against concepts of basic science. Once a concept is adapted it can not be violated, set aside so to speak, to overcome a difficult spot in the progression of the model. So it may happen that the need for conceptual coherence forces one to tell paleontologists, for example, that their interpretation of the fossil record is against our science of knowledge. Fossil hunters are not a humble breed and the danger that they will go into depression over this or anything else is very slim. Experimenters, however, are not automatically the best conceptualizers as we learned from Tycho Brahe, the Danish astronomer, who knew the stars better than anyone during his time, Kepler included, but remained a Ptolemaist until his end! This dichotomy between global and local views is quite obvious in every aspect of evolutionary sciences.

Concepts may have quite an independent life. Relativity was a purely theoretical entity before the first measurement confirmed the idea. Of course, we need experiments to solidify our mental constructs but we do not need experimental proof if there are no theories and these theories are not strewn about on any nature trail. One will not do experiments without motivation; the two processes are clearly distinct.

One origin versus billions of origins of life is the first and the crucial difference between Darwinism and the Genomic Potential Hypothesis and one wonders how, in this one and only world, such startling difference can be perceived. At least one Darwinian pointed out, with little echo, that one origin is impossible for physical chemical and biological reasons and upped the ante to at least ten,[3] still missing the point by a wide margin. The unlikely inspirational journey toward the Genomic Potential Hypothesis through the pages of chemistry texts provided images of mass action, $10^{23}$ particles reacting until $10^{23\text{-}X}$ were on one side and $10^{X}$ on the other. The starting molecules (reactants) and the conditions always gave rise to a main product and some by-products. Changing the conditions produces additional distributions of products. This is how biogenesis had to have occurred, large numbers of products within a matrix of by-products, and that is a truism for those who are convinced that chemistry is at the base of it all.

Connecting to reality one notes that experimenters have provided evidence for the presence of microorganisms in Hadean rocks. They were not set on providing evidence for the Genomic Potential Hypothesis because they did not know it. They are working for Darwin and so far their effort netted about 3000 organisms which have been identified to date as belonging to 300 different species, 90 of which have modern counterparts;[4, 5] just what one would expect from the mass action law of chemistry. The face value of these pictures is unmistakable; but do they really indicate massive production of first cells? At this period in the earth history, yes, there was not enough time between the appearance of so many cells and the cooling of the earth to allow for a single origin scheme of gene duplication and mutation. There is excitement when model and nature agree with each other, the excitement of one step forward. The argument continues that if 3000 microorganisms have been seen in 3.5 to 3.8 billion year old rocks perhaps 3 trillion or $3 \times 10^{12}$ were there so as to provide the chance for us to find 3000 during random searches 3.5 billion years later. It appears that the paper trail pays off!

All scientists agree that chemistry is the basis of life and with that preamble it seems almost certain that the first cells one sees in the Hadean stones are the first ones on earth; their ancestors were the heat- and light-driven bio-reactors. The Darwinians see the same cells but because the (in principle unprovable) single origin is a creed of the model, they postulate that a single ancestor must have lived much earlier and mutated into all of the cells that are visible at this horizon. At this point the old hypothesis is defeated by the researchers in planetary sciences who do not see any 'biology-time' before the time of the first massive invasion of the earth crust by cells. The hell fire of global accretion was too close for "descent with variation".

The evidence that needs to be incorporated into the different hypotheses is the same. Looking at a cross section of the Devonian period one sees many kinds of fishes, amphibians, nautiloids, and arthropods of all descriptions, dotting the space through which hypotheses can roam. Darwinians see them as the products of incessant branching while genomists see the blossoms of new clones, new shoots from the global tangle of the roots of life. Clearly, more fossils can deepen our understanding and will eventually help to put reality into focus by limiting our models, but the resolution of the polyphyletic vs. monophyletic dichotomy will depend not upon fossils, but upon the verdict from biological sciences as to the feasibility of inter-species conversion. If sharks appeared, as they did, within less than 10 million years between the middle and the upper Devonian period,[6] then one knows from science that these animals were derived from a shark-specific living system, and from the fossil record one knows by inference, that this ancestral system must have looked very much different from a shark. The Genomic Potential Hypothesis predicts that clones had transgressed into phenotypic existence in a few steps, from a quasi ovum or stem cell, and grown from miniscule to something visible in the fossil record such as a shark.

We need fossils and we need them badly. But discovery has many facets and it is imperative that the various aspects come together from different and independent efforts to average out an individual's bias; creators of models of nature may be great guides through the minefields of ideas but they are never good scouts for evidence. The perils are not so much poisonous snakes and yellow fever as the capacity for self-deception.

## References

1. Johanson D, Edey M. Lucy. New York: Simon and Schuster, 1981.
2. Kuhn T. The Structure of Scientific Revolutions. Chicago: University Press Chicago, 1970.
3. Raup DM, Valentine JW. Multiple origins of life. Proc Natl Acad Sci USA 1983; 80:2981.
4. Schopf JW. The evolution of the earliest cells. Sci Ami 1978:111.
5. Schopf JW. Microfossils of the early archean apex chert: New evidence of the antiquity of life. Science 1993; 260:640.
6. Stanley SM. Earth and Life Through Time. New York: Viking Press,1986.

# Chapter 4

# The Origin of Complexity

Clearly, molecular complexity was a precondition for biogenesis. Contiguous molecular structures are monkey bars for electrons that provide the stream of energy required to extend that complexity to a level not observed in the abiotic world and to constantly renew the framework of life. Electrons built these structures via overlapping orbitals but for that to happen one needs solvated molecules dancing about each other to the tune of thermal motion until they stick together as their tiny magnets click. Long structures, branches, triangles, pentagons fused to hexagons; quite a panorama. One would observe, if possible, that molecules sort themselves by structural fit, leaving out the bulk of molecules which, in turn, may associate in different affinity groups. Nucleotides were favored structures which were produced in large amounts because the monomers were removed from the equilibrium by polymerization to form long chain nucleic acids, and that constantly shifted the equilibrium to produce more monomers and so on, leapfrogging to the first plateau in biogenesis, the potential genomic material. So, it was not a desert and it was not an ocean of infinite dilution, it was a field of perfect opportunity with moisture and temperature cycles that supported primordial chemistry under the unique condition that an earthlike planet produces only once in its history.

After decades of dramatic advances in science could enough have remained unknown for so long as to give substance to a new global paradigm of evolution based upon a different view of what chemistry can cause and how? The point is that every thing needed for the new model is known, it was all there, but nobody examined it from a different perspective and put it together accordingly. Just looking at something from a different angle can indeed give rise to a totally different model. Every hypothesis in science has been conceived that way. Relativity is merely a different perspective of the universe that was Newtonian until 1915. The Genomic Potential Hypothesis is the result of viewing a Darwinian world from a chemist's perspective and the two models are at least as different as the old and new astronomy.

So, let us ask once more what caused organic material to assemble itself into those delicate structures within structures in multi-dimensional symmetry that form the skeleton of life? What pushed immortal equilibrium chemistry onto the precarious perch that nonequilibrium chemistry occupies? The answers must lie in the realm of chemistry for the simple reason that, when life assembled itself, there was nothing else. The organizing principle is chance in the old model while the new model points to atomic orbital steering effects; the difference lies in the predictions that arise from each model as exemplified by the single origin versus the multiple origins paradigm of life. Let me elaborate upon these rather basic observations.

If such an eternal phenomenon is indeed responsible for biota then one might suspect that earth and life could have come about at the same time. Well, it almost did! About 4.5 billion years ago the earth did provide the harsh milieu for uncatalyzed chemical reactions, which include temperature gradients, reduction potential, and the necessary elements. The escape of gas from the earth's interior and the reduction of carbon compounds to hydrocarbons were early events. More carbon was brought in by carbonaceous chondrites (carbon-containing meteorites) and water was probably added to the earth's surface in small increments by comets which are lumps of water ice (as opposed to ammonia- or $CO_2$-ice). The earth's atmosphere consisted of heavy gases such as carbon monoxide (CO), nitrogen ($N_2$), hydrogen cyanide (HCN), and methane ($CH_4$), that could not escape from the gravitational field. Free oxygen, however, was absent or minimal.[1-3] The stage was set for an interesting play, the actors were simple organic gases, and the hot rocks of the earth's surface were interactive props. Volcanoes added sulfur and the traces of water wherein organic compounds stewed for about three- to four hundred million years, or so goes the story that is quite familiar to the reader.

Still, all of this is chaos, a hodge-podge of molecules zooming about each other seemingly unmanageable, and that certainly invites the thoughts of an all-pervading organizing principle to which everybody agrees, albeit under different names. The evolution of this concept goes from the gods of religions to the chance of Darwinism to the atomic structure of genomism. Where then do the new Genomic Potential Hypothesis concepts enter the old picture? It happens at this very point that a genomist notes that the order is inherent and appears from within very subtly to give viscosity to the mixture. Carbon is the thixotropic agent, the atom that organizes the world around itself in five dimensions, one for each corner of the 4 $sp^3$ orbitals and time as the fifth.

This reading differs a little from the dimensions of physics, but viewed from the nucleus of a carbon atom it seems natural (Fig. 4.1). Hydrogen, nitrogen and oxygen have not been mentioned but if we admit them, 98% of the constituents of living systems are accounted for, it merely remains to activate them. The directionality and number of the orbitals is unique for each element, and all substituents not only take their position in space[4] but also dictate to a point what atoms may react at the remaining sites. Liberal as these limits are they select against millions of other interactions, with further selectivity introduced as the molecules get larger and by specific conditions of pressure, temperature, pH and oxido/reduction status. In fact, the chemistry of the origins of life was most likely dependent upon a gradient of conditions (c) dc, as a function of time (t) dt, beginning with hot organic syntheses of components and continuing with less energetic chemistry of polymerization; and lastly, the low temperature era of structure mediated catalysis which ended with the appearance of cells. This scene actually matches what earth science tells us about the conditions on the post-accretion earth.

Bonding orbital disposition and reaction conditions are all that is required, but what to do with that information? Can we take these conditions and properties to reconstruct in detail the biogenic path? No, and that does not mean that the concept is incorrect but rather that it is not knowable down to sufficient detail by human brainpower. It is not the complexity of data but rather the number of constants and the miniscule difference between them that prevents us from developing life from first principle. Even if we were able to solve the Schrödinger equation it would not get us

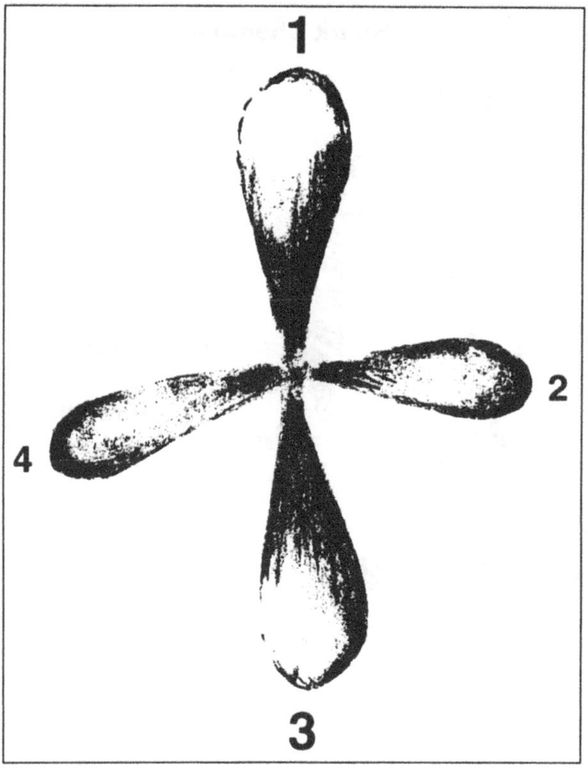


Fig. 4.1. The four lobes of the carbon atom evolving from the nucleus are probability distribution ranges of the bonding electrons. Lobes 1 and 3 project above the plane of the page and lobes 2 and 4 below. Thus carbon organizes the world around it in four directions with time as the inevitable fifth component. A chain of carbon atoms will not make a straight line but rather produce a continuously changing structure in solution, determined by defined angles and distances. Reactivity of the atom is restricted to the four tetrahedral bonding orbitals of carbon, and thus all in between orbital bonding is not possible and that is what has been referred to as loading the dice. Only a few degrees difference in the bonding orbital disposition would have made the development of life impossible.

very far into biogenesis;[5] physics is too simple to explain life beyond its atomic basis! Mono-and bimolecular reactions are fairly predictable, but to see where, in very complex reaction mixtures, the various atoms end up one must surrender to empiricism. A number of years ago an intriguing experiment was performed by Stanley Miller.[6] Regular laboratory glassware was used to assemble a closed system fitted with in- and outlets, an electric spark gap, a bottle full of water and a heat source. Oxygen was removed from this system and the same "primordial" gases were entered that exist in the interstellar space.

The water in the closed system was brought to a boil and the vapors, rising through a discharge chamber, were condensed and led back into the reservoir. After a few days of recycling, the apparatus was opened and the contents analyzed. The result was so amazing that more and more proof was demanded by the editors, but when the amino acid glycine was crystallized from the mixture there was no escape, Dr. Miller had produced amino acids and the bases of nucleic acids from primordial gases by a random process! At least that is the prevalent interpretation today.

The genomist asserts that this experiment has shown beyond reasonable doubt that chemistry is not a random process! The reader, chemist or not, can verify the genomist's assertion by counting the atoms in the mixture, (assuming that they are spheres without features) and do a simple probability calculation. The result shows that amino acids and purines and pyrimidines should have occurred at such a low concentration that they would not be detectable with our technology. Carbon, nitro-

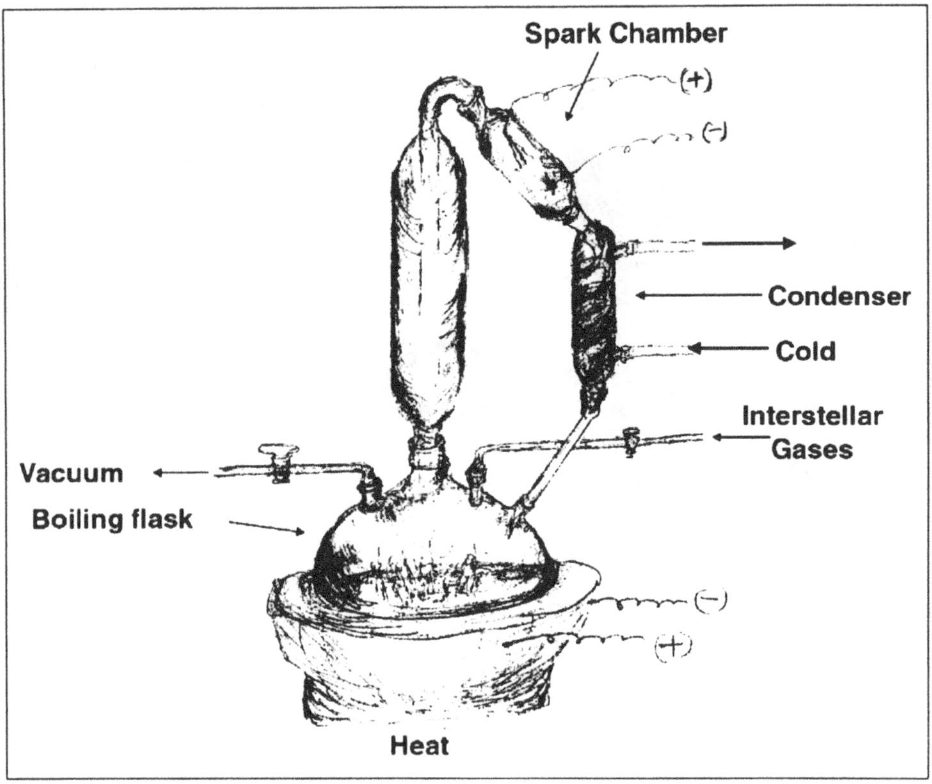


Fig. 4.2. The "Miller-Urey" apparatus for abiotic synthesis of biochemicals from primordial gases is shown. Before each experiment the system was thoroughly evacuated, flushed with interstellar-type gases, and sealed. Water is brought to a boil and vapors rise through an electric discharge chamber and are re-condensed and led back into the boiling water reservoir. It took only a few weeks to produce a color change in the water which indicated an accumulation of organic compounds shown in Table 4.1. On the young earth, of course, this experiment would have been carried on for a few million years.

gen, oxygen and hydrogen would never have come together as they did if reactivity would depend only upon the sequence in which molecules collide with each other. Important is the sequence in which they stick to each other and how tightly! Of course, molecules have to collide before they can stick, but the number of unsuccessful collisions is very large compared to the number of successful ones because of steric steering. The significance of this experiment reaches far beyond the demonstration of abiotic amino acid and nucleotide production; it gave evidence for the self-organizing principle that lies at the root of evolution. Note how far we have strayed from the Darwinian path by just taking a fresh look at the old problem. The result of the Miller/Urey experiment has been greeted as curious or remarkable but not as what it seems (to me), i.e., the discovery of the "background radiation" of the birth of biology. So much for the difference in perspective! Although Miller's findings went far beyond Darwinism, he remained a Darwinist until today, and that is astounding testimony to the unwillingness to take ever so small a step away from the polished surface of an experiment, and to dare to be distracted by its meaning in second and third intention.

The "Miller-Urey" apparatus and the primordial puddles have a number of properties in common, namely:

1. Sterility,
2. No free oxygen,
3. Primordial gases,
4. Small amounts of recycling water,
5. Energy, and
6. A mineral surface.

The common argument that the ground will be different in different regions is countered by the fact that results of similar experiments done by nature under significantly different conditions in the asteroid belt have led to nearly the same result. The carbon compounds isolated from carbonaceous chondrites (meteorites) are the ones also seen in the laboratory experiments. What an impressive demonstration of the principle that is to be highlighted in this chapter. The Miller experiment made one recognize that all that was required to produce bio-molecules of great variety was to activate confined, relatively inert gases; the structure of atoms did the rest.

The experiment makes yet another exceedingly important point, namely that, no matter where such a reaction would have been set up and how activation was affected, the result would always be the same. This conclusion is supported by the data in Table 4.1 which lists the results of the analysis of a natural experiment done in the asteroid belt as compared to the products of a Miller experiment. In as much as our galaxy is not notably different from others, one may expect the organizing force of chemistry to be indeed a cosmic phenomenon as is life. Continuing this train of thought, are there still difficulties in imagining that organic puddles in Texas would yield chemical products similar those located in what is now New England?

The structure of atoms is known to us in great detail down to the fact that isotopes are not totally equal to the major form of an element. Heavier isotopes react slower in enzyme-catalyzed reactions but they form stronger bonds. No uncertainty is needed for that aspect of our world. Heisenberg's Principle does not pertain to the power fields created by the atomic nuclear structure, but rather to the position of the electrons within these power fields. For the macroscopic world and its chemical basis the position of the power fields (orbitals) is important, and these positions allow carbon, very precisely and very predictably, to give rise to four bonds (or two double bonds as in $CO_2$ or a triple and a single bond as in cyanate) that are stable under a variety of conditions and which allow carbon to form the many and varied polymers that form the skeleton of life. Chemistry does not suffer uncertainty neuroses.

The doyens of physics, Einstein, Planck, Schrödinger, and even Heisenberg, have warned against a view of the macroscopic world based upon subatomic uncertainty. Einstein, in a dialog with Murphy, calls it "not just nonsense but objectionable nonsense".[7] To no avail, once the principle was announced indeterminism became the patron saint of human dignity, superiority and free will, and luck became a legitimate adjunct to science. Karl Popper once called Darwinism an unfalsifiable hypothesis, i.e., not a scientific hypothesis until the indeterminate wave in London swept him off his pedestal into the gully of politically correct thought; he was no Giordano Bruno.

As it is, almost 99% of the human body consists of hydrogen, carbon, nitrogen, and oxygen, which are the members of the first and second period of our table of elements. It is perhaps easy to realize that the order of elements in the table reflects the

***Table 4.1 . Relative abundance of amino acids in the Murchison meteorite and in an electric discharge synthesis***

| Amino acid | Murchison meteorite[a] | Electric discharge |
|---|---|---|
| Glycine | **** | **** |
| Alanine | **** | **** |
| α-Amino-n-butyric acid | *** | **** |
| α-Aminoisobutyric acid | **** | ** |
| Valine | *** | ** |
| Norvaline | *** | *** |
| Isovaline | ** | ** |
| Proline | *** | * |
| Pipecolic acid | * | * |
| Aspartic acid | *** | *** |
| Glutamic acid | *** | ** |
| β-Alanine | ** | ** |
| β-Amino-n-butyric acid | * | * |
| β-Aminoisobutyric acid | * | * |
| γ-Aminobutyric acid | * | ** |
| Sarcosine | ** | *** |
| N-Ethylglycine | ** | *** |
| N-Methylalanine | ** | ** |

Data, Miller 1974 *Mole ratios to glycine (= 100): *0.05-0.5; **0.5-5.0; ***5-50; **** > 50.

same natural law that forms the elements in the first place, and it would be logical therefore to search for compatible properties among neighboring elements. What our elements of life have in common is smallness and the ability to form stable bonds by adding either one, two, three, or four electrons to their outer valence shells. From the third period of the table of elements only phosphorus (P) and sulfur (S) are important for living systems because of their specific electronic structure. Again, it is the specific atomic configuration that allows both of them to have structural roles as well as energy-transferring roles. The backbone of the genetic material (DNA and RNA) contains phosphorus, and protein cross-links contain sulfur. In addition, phosphorus is used as "currency" in biological energy transfers. The argument has been made by Edsall[8] and later by Wald,[9] but the message was lost on biologists.

Everything seems to be logical in the context of chemistry. The important atomic characters can be displayed such as to reveal the reasons for self-association in an orderly fashion. From a purely mechanistic point of view we need something that will cause inanimate matter to stick together to produce, over and over again, similar structures and variations thereof. Figure 4.3 shows a bare skeleton display of molecules which come together at specific angles at very specific distances because of the bond angles of carbon, nitrogen, and oxygen. Molecules are a little like an erector set, standard distances marked in standard components. The structural information of atomic orbitals is propagated to the molecular level.

The corners of these stick models are carbon and the lines leading away from the corners are bonding orbitals. Dimers, trimers, and tetramers come together, and when

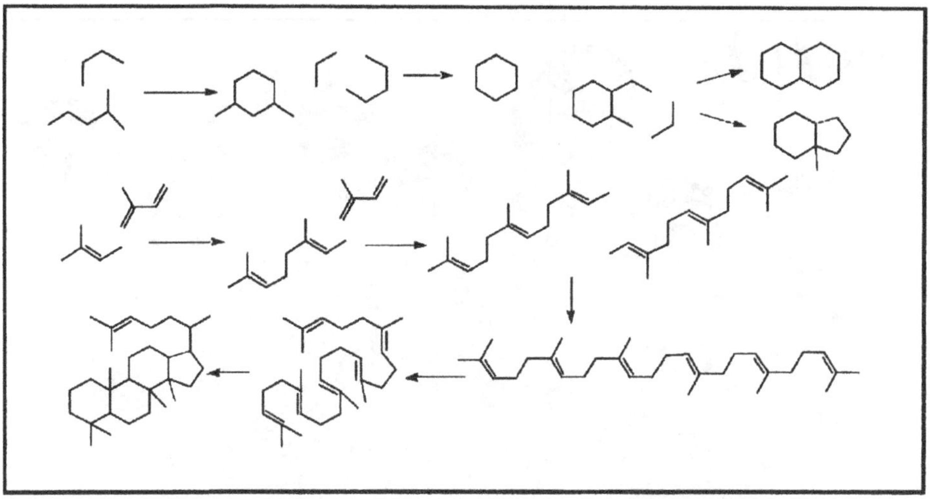

Fig. 4.3. This figure illustrates the "erector set" quality of carbon chemistry. Carbon-carbon bonds are of unit length and produce unit angles so that electron activation will cause repetitive, spatially defined structures that, under appropriate conditions, will lead to self-reproducing surfaces.

they align their ends they fit precisely because the carbon-carbon single, double, or triple bonds are precisely defined to give the building blocks for molecules a natural fit. It stands to reason that monomers produced in that fashion will again have functional groups displayed such as to overlap the neighboring molecules and cause the buildup of superstructures. In Fig. 4.4 the same stick models are drawn with their electronic clouds to show how chemistry might look if our eyes were sensitive to x-rays.[10]

Clearly, it is a mass-action phenomenon and the only selection that can occur in such a solution is the fit of specific groups to build larger molecules. In this way the bonding orbitals, the direction and strength as well as their sensitivity to competing reactions, cause the buildup of more complex molecules if reaction conditions favor such a development. In Figure 4.5 the concept is illustrated by an actual example of a radiation-driven, nonbiological series of reactions that ends in the formation of proto-porphyrin.

These pictures[11] are simplified to provide an unobstructed view of one of the most important principles of the inanimate world. The power of the Genomic Potential Hypothesis stems from the realization that there is no purpose and no goal in all of this and that syntheses came about because of the predisposition of atomic and molecular structures for such reactions under certain conditions. In contrast to the chance-oriented Darwinian paradigm, this model invites experimental exploration.

Prebiotic self-association was inevitable as well as uniform at the most fundamental level; differences are the consequence of higher level organization, i.e., the order rather than the ingredient. Once a nucleic acid sequence had been established and became part of the memory of an organism, it was maintained by the mode of complimentary self reproduction for as long as the species lasted. Again it is the uniform distance of hydrogen bonds that line up the nucleotide bases for accurate reproduction.

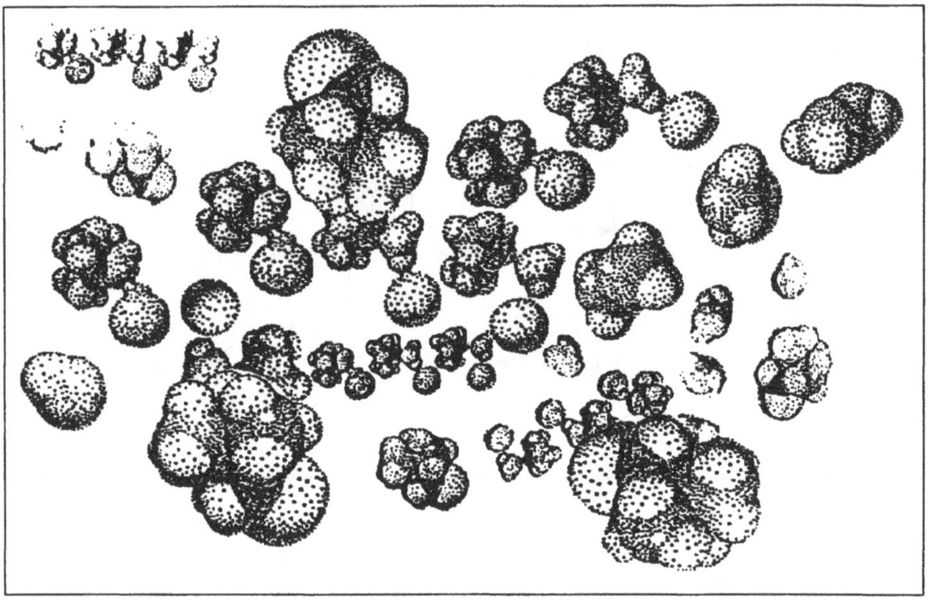

Fig. 4.4. This figure may be considered a window to a "chemical aquarium." The core of each of the structures is carbon and most of the hemispherical structures attached to the core represent hydrogen atoms. When these molecules collide under appropriate conditions, either on the surface of a catalyst or with sufficient energy, the carbon core of two units will form covalent bonds, displacing hydrogen. The discharge chamber in the Miller-Urey experiment served to displace the hydrogen and thus to create active molecular species that would form larger covalent structures.

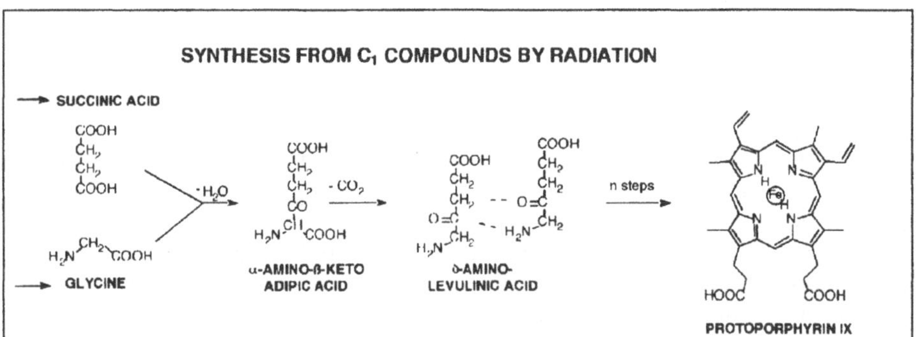


Fig. 4.5. From the mixture in Figure 4.4 a chain of reactions has been isolated to show how the guiding force inherent in the electrical orbital orientation can lead to complex bio-molecules.

Note how many distances have to be uniform in order to properly align the hydrogen bonds that stabilize the DNA helix. Here the principle of bonding orbitals is continued to the next higher level of molecular association. The next step, the covalent linkage of the bases of nucleic acid to polynucleotides, is the moment when potential memory is produced from molecules that singularly have no meaning.

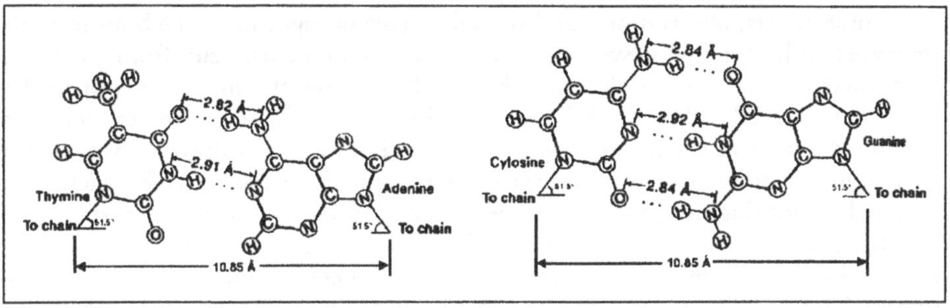


Fig. 4.6. Here the principle of bond length and angles is shown in connection with the core molecules of all of life, the monomers of DNA. The configuration of bonding orbitals leads to the selection of bonding partners in the DNA and this arrangement again causes the continual reproduction of DNA through the complementarity principle. While this concept applies to molecular interactions in general, there is nowhere a more impressive demonstration of a principle that is central to life and central to the Genomic Potential Hypothesis than this threshold of information production.

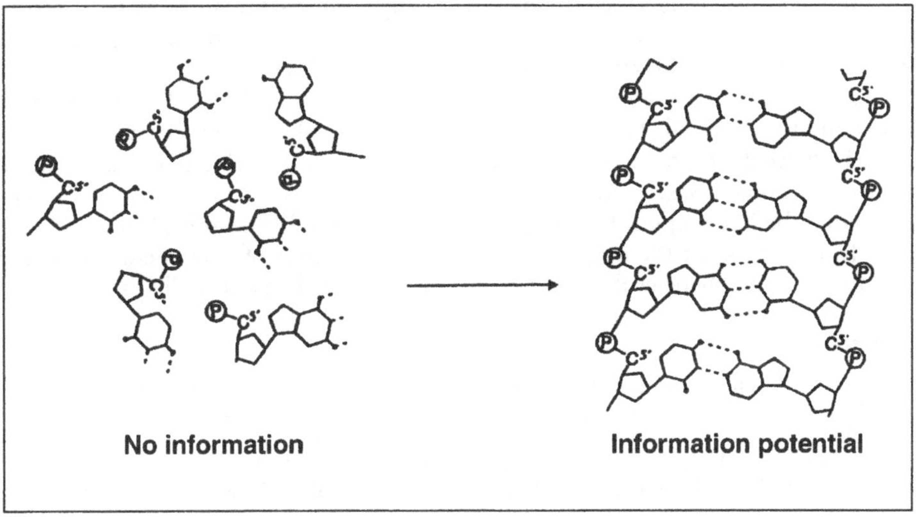


Fig. 4.7. This figure illustrates the chemistry of memory. The process is quite well known, but in the new hypothesis it is given the attention it deserves. The monomers of DNA designated as containing "no information," of course, do have the information for self-assembly. The hydrogen bonds sticking out from the single nucleotides are the signal for alignment on the primary strand for negative/positive reproduction of the genomic material as well as for the mRNA that translates the information into protein sequences. As a unit, the DNA holds information for the organization of chemistry along its length, which eventually translates into organisms.

We do not know precisely how it happened, but we know it did. Would that necessarily stamp evolution per se as a soft science? Not really. Or would one move physics into that category because among other things, the origin of inertia is still not known? Every living science must have its unknowns because, once all is known (which is anyway impossible) the discipline becomes history. The foundation of the Genomic Potential Hypothesis lies in chemistry, which is governed by thermodynamics, kinetics, and the laws of mass action; nothing soft about that!

These events, observations, and thoughts have brought us to the bottom of the staircase that leads to living systems. The production of nucleic acids from nucleotide monomers is not totally understood, but we know how the single units of nucleic acids are produced and we know how nucleotides need to be connected to each other to form the basis of the bio-memory (Fig. 4.7). Formation of these high-molecular weight, linear molecules was the crucial event that provided a path to life. Nucleotides changed to nothing new, they just joined hands (orbitals) to form the surface from which, by a circuitous route, other complex molecules could be read. Again, we do not know the details of the abiotic polymerization process for nucleic acids, but it had to have happened by fusing bonding orbitals guided by steric restrictions. Someday our studies of the chemistry of the origin of life will present us with the mechanism of abiotic polymerization of nucleotides.[12] Complexity hereafter is defined by how many times and through how many levels of structure molecules are connected. The surfaces produced led to the reactions that increased the speed of growth of these structures and provided the energy for maintenance, and when all was wrapped up in a membrane the border was crossed to that "mysterious state" of life without putting anything mysterious into the reaction. The change from inanimate to living systems is not so much a change of basic materials, i.e., carbon, hydrogen, nitrogen, oxygen, but rather the change in the way they are connected to each other.

The structure of energy makes the emergence of life inevitable if the proper reaction conditions prevail. This, the determinist's clarion always sounds true because its score is chiseled into the structure of energy manifest in atoms. It also is a useful sound for it tells us that, if we should learn how to get to another earthlike planet, our test for life should be based upon the very principles that led to life on this planet.

To the genomist the origin of complexity is a matter of coincidence of natural conditions and, wherever those conditions are met, life like ours will arise.

## References

1. Ferris JP. The chemistry of life's origin. Chem Eng News 1984; August 27.
2. Canuto VM, Levine JS, Augustsson TR et al. The young Sun and the atmosphere and photochemistry of the early earth. Nature 1983; 305:281.
3. Rebek J. Synthetic self-replicating molecules. Sci Am 1994:48.
4. van't Hoff JH. Imagination in Science. Berlin-Heidelberg-New York: Springer Verlag 1967.
5. Schrödinger E. Naturwissenschaften 1926; 14:664.
6. Miller SL. The atmosphere of the primitive earth and the prebiotic synthesis of amino acids. Orig Life 1974; 5:139.
7. Planck M. Where is Science Going? Woodbridge: Ox Bow Press, 1981.
8. Edsall JT, Wyman J. Biophysical Chemistry. Vol. 1. New York: Academic Press, 1958.
9. Wald G. Life in the second and third periods. In: Pullman K, ed. Horizons in Biochemistry. New York: Academic Press, 1962.
10. Pauling L. Chemistry. Sci Am 1950; 182:60.
11. Calvin M. Photosynthesis and quantum conversion. In: Pullman K, ed. Horizons in Biochemistry. New York: Academic Press, 1962.
12. Ferris PJ, Hill AR, Lui R et al. Synthesis of long prebiotic oligomers on mineral surfaces. Nature 1996; 381:59.

# Chapter 5

# Our Young Planet:
## One Is Not a Choice

The old model begins with one origin and the biochemist knows that to be impossible. The scientific basis for the genomist's position has been discussed and will be on the agenda again, but now it is time to pay tribute to the plausibility side of the argument. The approach taken in this chapter will, better than any other I can think of, demonstrate how natural it is to think of millions of origins of life.

Choices can be limited for many reasons. There used to be a potato chip company that declared it impossible to limit oneself to eating just one out of a full bag which one knows is not true. In chemistry, however, there are conditions where "one" is truly not a choice. 'One' does not become a significant number until the arrival of cell biology with restraining membranes. The chemistry of life's origin is based upon mass action effects and can never be limited to one outcome! This is the stark, but not necessarily unpleasant, reality that settles the fate of the Darwinian paradigm. Allow me to take you on an imaginary trip to the Hadean period, the time and place where the problem comes to life, back to the stage of the reiterative polymer chemistry.

Scenarios are reality checks for perpetrators of new ideas. It makes a poor impression if a story, as it runs, suddenly requires a miracle to continue on course. Visiting the past is in principle and in reality possible (astronomers do it routinely) and all studies of our evolutionary history are attempts to do so based upon a mixture of understanding and prejudices. One is not seeking truth on such a trip but rather plausibility for a concept that must be true by the science of it.

Let us wing our way down to the fantasy surface of the early earth where the first larger cratons of silica have formed.[1, 2] The cratons are sitting like gigantic mesas in an expanse of flat and low-lying land[3] from which smoke is rising at various places like one imagines the signal fires of our native Indians to have dotted the North American prairie. It must be the year 600,000,000 after accretion (AAc). Small dimples of moisture exist under the caked surface of the desolate place of our beginning. It is too late in the history of the earth to observe the chemical synthesis in full swing and it would have been, of course, much hotter at that time. There are no clouds but smog is drifting across the ground and it feels hot and steamy. Marching across the plain one would off and on hear the whistling sound of projectiles crossing the sky. At times the earth is shaking from an impact; not too far away a wall of water is rising, turning into a mist and drifting away in a gust of thermal convection. A meteorite impact, one of the many that homed in on different places several times a day.

Small depressions hidden under a carst-like surface are moist. Pushing a finger briefly into the ground actually produces a few drops of water before the impression fills in. A pocket spectrograph shows a strong absorption line at 260 nm. As the samples cool in the instrument, the absorption drops measurably, suggesting a stacking effect, hypochromicity due to secondary structure which brings to mind helical nucleic acids. But how could this be, after a pretty hot accretion period that produced the iron core and molten rocks there should be no organic carbon save, perhaps, some methane from carbon reduction, but here we notice nucleic acid chains!

The instrument confirms this suspicion; the little depressions of moisture contain all the molecules we know from the science literature, nothing seems unexpected except that the earth looks arid. There are no balmy oceans and no water, only a little moisture in spots, certainly nothing drinkable. This is the time when a dreaming chemist would scour the earth for biogenic systems; flashing an ocean into his dream would give him a kinetic nightmare.

Counting spots (a human trait) of moisture one might see a fireball overhead, like a rocket grazing the edge of the craton, and continuing its plunge straight through the bottom of the low-lying plane. Molten rock wells up as the crater closes. The crust in the future ocean basin is fairly thin, the ground is shaking like Jell-O, and the shock wave propagation could be felt for long distances. All the little reservoirs squirted water like miniature fountains as the compression wave propagated, and for a few minutes a shallow flooding was visible which disappeared like a mirage during the next wave of expansion, sucked up as it were by the porous ground. A dust cloud rose from the impact side high into the atmosphere, some of it settled at once and was washed into the ground as another oscillation squeezed and reexpanded the surface. All the puddles in this area look alike, clearly a case of impact mixing. On higher ground certainly the composition would be different but here again a surprise, the basic structure of polymers is all the same except that here and there one would detect more amino acids, some fatty acids, and some alcohols. The hypochromicity at 260 nm was more or less pronounced in some of the areas but qualitatively the spectra were alike.

If this seems amazing to the reader one might consider once more the Miller-Urey experiments which have been set up everywhere on earth where scientists are investigating the origin of life.[4] The results reported from different locations were the same if the conditions were the same as well.[5] Remembering the meteorite analyses we can state that if one had an unending string of Miller reactors stretching through the universe one would find the same compounds in all of them. Changing the initial gas mixture in one of two reactors side by side in the laboratory would yield different product preponderance, i.e., mainly amino acids or purines and pyrimidines and so on.[6]

Back at the scene we are examining the puddles with an instrument that sees nucleic acids only and thus the pools are either positive or not for nucleic acids; from our science today we know that this means biogenic potential or not. Up to this point multiplicity is the rule and barely noteworthy, but the pressure of mass action laws would make singularity a miracle. All puddles looked the same and only occasionally one with a very high sulfur concentration or a high concentration of minerals was spotted which did not seem to carry the signature of nucleic acids. Larger bodies of water did not show any signs of nucleic acids and one became impressed by the observation that the makeup of the ground was less important than the total volume of water; small was better and more productive. Thoughts meander along to higher ground. Looking back the whole field of puddles seemed to lie in a plane surrounded by an elevated rim. Apparently a huge impact crater, a well-defined era for biogenesis on this craton. Suddenly the sky picks up a strange glow as a kilometer-size rock quietly makes its way right to the center of this valley, sending rocks and sand half-way across the craton; a screeching sound and a thunderous explosion come like an afterthought. Nucleic acid would get a ride on these particles to produce the first mass infection of half a continent. Beyond the reach of this newly discovered spreading mechanism there would be no nucleic acids. For completeness sake one would test puddles well beyond the reach of the impact and find the same materials that were seen in the valley of biogenesis. Perhaps the splash was not necessary to spread primordial chemistry; anyway, it was a good idea while it lasted.

While standing there contemplating the apparent inability to extrapolate natural phenomena, a whole gaggle of icy comets hit the low ground, one after another, for several hours, and when it was over the first sauna on earth had been produced. Shallow flooding remained for a few hours until the water had evaporated again or drained underground. In fact, in the distance one might see a little silvery reflection on the future ocean floor; water! Water and carbon, it seemed, came by the trainload from space and mixing of primordial chemicals was a widespread occurrence.

To continue the thoughts that were interrupted by the comets as to why this unpredictability. Examining the nucleic acids from different places, similarity seemed extensive even between samples that came from widely separate spots. There are only four main components in addition to odd occasional nucleotides and that did not make for much variation, and the idea offered itself that life would eventually be read from any nucleic acid by cut and paste processes in line with energy minima. That might be the origin of exon/intron structures in eukaryotes.[7]

The 20th century had glorified itself by whimsical propositions such as assigning the origin of life to a one-time lucky event, which could have been that creative event over which evolutionists faced off with the creationists in court.[8] The evolutionists won, which was good, still it was a gift. In fact, it is a tragedy in my perception that lawyers had to decide that evolution should be preferred over creationism. Laxity in conceptualization must not be allowed to haunt us through the next century; there is so much to be learned. Thus this trip to the post-accretion world was organized in order to expedite and redirect progress into the new dimension.

It is now 800,000,000 years after accretion and the little puddles had a smell different from that noticed during the previous visit. Hypochromicity was barely observable but life could not be seen either. This would be the time to check out H. Huxley's Urschleim, *Bathybius haeckelii,* which was the creation primarily of imagination, supported by an artifact produced inadvertently by shelf-drying of samples dredged from the seafloor.[9] It had to be an artifact because the pan-protoplasmatic form of life (there are at least 12 names for this type of pro-form associated with prominent names in evolutionary history) had disappeared with the end of the biogenic period to form all of the micro-organisms that we find in Hadean rocks.

Miles and miles across the craton it became apparent that, wherever water or moisture appeared, a thin film of presumably organic nature could be found more or less pronounced. Urschleim? No, the osmometer reacted wildly to the slippery material, there must be membranes. The clones and subclones of cells are growing up at various distances that were similar even beyond the reaches of the many mechanisms of mixing.

Had we missed the time of biogenesis? The analyzer showed the release of gas from the slime layer on the rocks, but also from the surrounding water. A close-up view of the few milliliters of water here and there showed membrane fragments floating around, wrapping other matter into its confines and closing up and breaking open again, uniting with other pieces until a few stayed together and eventually joined the layers attached to the rocks. After some time it became clear that any unit which persisted for a few minutes was stable for as long as one could watch. It looked like a spare parts exchange station: I have this, you have that, let's get together. If it is insufficient, osmosis will open the unit and the trial and error search for complementary pieces starts over again. Some of the nucleic acid polymers produced protein in the viscous matrix that floated through the minuscule portion of a small reservoir. One could observe chains of proteins peel, stop and go, from a ball of nucleic acid (one could not distinguish RNA and DNA chains) perhaps one chain a day or so. The nucleic acid embedded in a matrix produced different proteinaceous products seemingly for nothing. Sometimes two or more of these core particles would coalesce and the extrusion of proteins would stop. A few minutes later gas bubbles would evolve from the newly formed globule. Methane of course, Archea![10] The formation of primitive archebacteria had happened right before our eyes. For the fun of it only, it could be nothing else, the gas analyzer was pointed at the object and watched as it pointed to oxygen! But oxygen evolution would require a very complicated molecule, a light-capturing electron transferring complex[11] that must have been synthesized outside of cellular confines, i.e., proteins without cells. The end of the long story is that every one of the cells in this pool really emitted oxygen and the slimy stuff did as well. Pools half across the craton did. Cyano bacteria! Blue-green algae, the first energy concentrators had self-assembled. Soon they would be eaten by those who had no talent, thus establishing early on the order in this world.

The story was amazing but of serene simplicity. Cells were forming from nuclear particles that had a variety of catalytic capabilities and, within a few seconds after a successful combination of particles had occurred, there was a living cell producing oxygen. All the proteins needed for these first steps had to be prepared and ready before the cellular structure could assemble itself successfully. That seemed to be the answer to the puzzle, i.e., the creation of living units within a split second whereas the total process or the period of biogenesis may have taken millions of years. Life was sparked into existence unit by unit until, at some point, the raw material could no longer exist because the cells took it up as a readily available building material and thus the process of biogenesis became too slow to compete.

Nice story, but could one assume that even in areas where mixing did not occur the photosynthetic apparatus, among other functions, should come into existence through nonequilibrium chemistry and be completely alike? With minor variations, yes. The structure and the function, i.e., photon capture, was preserved by necessity.[12]

In light of this 'eye witness' report, we must modify the cell-protein axiom to read: there are no cells without proteins and there are no proteins without nuclear material. Nuclear material was first and all the structural developments and inventions had to be made in the nucleic acid language, and that means it was made without a plan to create a specific activity but rather with enough variety that specific activities would arise among others. This concept should pertain throughout all life forms no matter how complex they would eventually be. The mechanism should lead to cells with redundant and useless proteins and even with useless functions, and gene-silencing (knock-out) technology in mice has delivered evidence to that effect.[13]

The Hadean may not be a first choice for a vacation, but for evolution the edge condition proffers an unequaled opportunity to gain new perspectives.

## References

1. Hutchison R. The formation of the earth. Nature 1974; 250:556.
2. Suess HE. Remarks on the chemical conditions on the surface of the primitive earth and the probability of the evolution of life. Orig Life 1975; 6:9.
3. Stanley SM. Earth and Life Through Time. New York: Viking Press, 1986.
4. Miller SL. The atmosphere of the primitive earth and the prebiotic synthesis of amino acids. Orig Life 1974; 5:139.
5. Mizutani H, Mikuni H, Takahasi M et al. Study on the photochemical reaction of HCN and its polymer products relating to primary chemical evolution. Orig Life 1975; 6:513.
6. Barak I, Bar-Nun A. The mechanisms of amino acid synthesis by high temperature shock-waves. Orig Life 1975; 6:483.
7. Dorit RL, Schoenbach L, Gilbert W. How big is the universe of exons? Science 1990; 250:1377.
8. Benton MJ. Diversification and extinction in the history of life. Science 1995; 268:52.
9. Welch GR. T. H. Huxley and the "protoplasmic theory of life": 100 years later. TIBS 1995; 20:481.
10. Pool R. Pushing the envelope of life. Science 1990; 247:158.
11. Deisenhofer J, Michel H. The photosynthetic reaction center from the purple bacterium *Rhodopseudomonas viridis*. Science 1989; 245:1463.
12. Barber J. Photosynthetic reaction centres: A common link. TIBS 1987:321.
13. Hooper C. Genetic knockouts: Surprises, lessons, and dreams. J NIH Res 1992; 4:34.

# CHAPTER 6

# The Condensation of Life

The course of biogenesis was carved into molecular structures by the events of the primeval initiation of our universe. The speed at which pure energy segregated into baryonic matter leads one to think that there was no other possibility. (Physicists quipped that God may not have had a choice in the matter). The Genomic Potential Hypothesis extends this inevitability concept into the biogenic events, and the speed at which life assembled itself from molecules immediately (by geological standards) after the earth had become stable enough for chemical processes certainly invites the conclusion that luck had no part in this matter either.

Darwinians view the origin of life as a lucky strike.[1, 2] Chance events, however, are irreducible and irreproducible so that comparison between the old and the new model becomes possible only from the moment when life had been established. Of course, every one invokes chemistry when it comes to the origin of life but for chemistry the 'single' is out of character. Nonetheless, the consequences of any origin of life scenario should be expressed in the fossil record and this is the part where comparison of models becomes possible. In as much as the Darwinists deal with the topic of this chapter with one word (chance), the Genomic Potential Hypothesis is alone in its effort to build a conceptual basis for biogenesis.

In the new hypothesis, biogenesis is a series of processes that must be justified by the basic rules of the relevant scientific discipline, chemistry, and must lead to verifiable predictions. The genomist predicts that innumerable origins evolved at many places on earth and that chemistry was the rectifier that caused uniformity at basic levels of biochemistry up to and including the genomic code. Thus, it is the burden of the genomist to replace the chance events of the old model with a series of principally known reactions of predictable consequences.

Every component for cell assembly has to be there in large numbers at many places and functional when the individual units slide into the state of life. These 'multi-origin' ideas have been around[3-6] but did not fall on fertile ground, and what remained of these mavericks' polyphyletic thoughts was wiped out by the discovery of the universality of the genetic code and the erroneous conclusions drawn from that observation.

We know with certainty that life condensed around nucleic acids because only nucleic acid remembers and transmits. Proteins are unstable and while they carry a lot of information, they do not reproduce. Only when the nucleic acid persisted from which proteins were made continuously could they become a factor in the equation of biogenesis. The nucleic acid buildup happened of necessity without a guide other than molecular properties and without a target.

The composition of nucleotide polymers at any locale depended upon the relative concentration of the nucleosides G, C, A or T, in the medium and may have varied as a function of time, ionic environment and other factors. The first chains that polymerized without a template produced variety whereas subsequently chains were duplicated more or less accurately by complementarity, and that process gave rise to families of nucleic acid polymers that formed the basis for similarity (clones) and diversity among organisms (Chapter 4).

Concepts of the ascent to life must follow a plausible path that will lead to a point where experimental verification becomes a prospect. Experience from the laboratory tells us that some of the key preparatory steps that involve nucleic acids and proteins do not require enclosure into cells in order to function. In fact, it was imperative for the biogenic process that the most important constituents of life could be produced by equilibrium chemistry, nudged on by reducing conditions and diffuse energy up to the moment when enclosure could be successful. Template-directed in vitro protein synthesis is quite well known to us and it seems reasonable that without an equivalent process in geo the primordial scene would not have worked. Cells must function in the moment they form and not a minute later. Heterogeneous reaction centers, the primordial aggregates composed of nucleic acids and the first few proteins produced by them, did not live but could persist, jump-skipping through their small puddle, until they met a complementary unit, one that had useful energy as an end product. These units were numerous in small volumes of liquid where they had been produced by mass-action from the same concentrated mixture of molecules. One must picture this process as one that provides a stage for many thousand years of progress and retreat, interspersed off and on by a success, i.e., a living unit produced in a few seconds. A few million years past the end of such a biogenic period an abundance of imprints of different fossilized cells should be visible in the Archean rocks and that appears to be true.[7]

Polymerization of nucleotides is an imperative for the story to come off. Clearly, catalysis by proteinaceous enzymes is a higher level complexity that had to await completion of the nucleic acid polymers. The recent discovery of nucleic acid catalysis has built a bridge between the nucleic acid and the protein world[8] and simultaneously cracked open the door to the multiple origin world. Nucleic acid catalysts were not as versatile and efficient as proteineaceous ones, but catalysts nonetheless.

The most important postulate of the Genomic Potential Hypothesis is that life was an inevitable consequence of the structure-energy manifest in atoms and molecules. Thus we have to address the question as to how one would build up something similar in different places when the nature of the product becomes apparent only many steps later? Obviously, words like goals or adaptation have no meaning in this setting. It would have happened only if there were limited possibilities, if energy and kinetics favored a set of reactions. Since the potential memory everywhere reads [(A)x (T)x (C)y (G)y], the nucleic acids in widely separated origins cannot look much different. But this "memory bank ", why was it so important for the progression of biogenesis? Because complex structures, other than the memory itself, that have not been read from the surface of a nucleic acid, cannot evolve. A chance assembly of amino acids, for example, will soon be lost when thermal motion has shaken it apart.

The message for assembling life in three dimensions is two-dimensional, meaning that the information it contains is sequence-dependent, and since the nucleic acid polymers in all primordial reactors would have different sequences the new model would lead to an infinite number of different life-forms. Within each pool, however, there would be important equalizing factors that would cause main themes and variants to arise from each biogenic center (which may have been as small as a liter). Once primary strands had self-assembled, complimentary strands would be produced faster than new primary ones, so that themes (primary strand copies) and variations (more or less accurately copied secondary and subsequent strands) could, for example, account for the variations that caused us to group species as families and superfamilies. That would put local order into potential chaos and would cause a viable mass of like organisms to occur and give each future species a reasonable base (number of members) for survival.

Global order is produced by the fact that life per se has a set of requirements such as energy, reproduction, and gas exchange that must be met regardless of the phenotype that a genome might produce, such as a worm, an insect, a mammal or a bird. From all the possible genomic forms only those succeeded that fulfilled these requirements, and so it happens that a great deal of similarity is observed among the catalysts of similar reactions in each life-form. The bacterial protein synthesizing machinery, for example, will work in mammalian systems, and hormones from tunicates and salmon perform specific functions in humans.

The crucial event for a multiple origin paradigm is the development of the genetic code. The only acceptable path to a genetic code for any model of evolution is via a structure/function relation between the main actors of the scene, i.e., nucleic acids and amino acids. RNA chains of varying lengths might have interacted with the free amino acids that were produced by primordial conditions and a cavity of sequence-dependent geometry could have been formed by the coiled-up RNA into which an amino acid side chain of L-configuration would fit.[9-12] These attempts to find evidence were not totally successful, possibly perhaps because the code-producing nucleic acid/ amino acid match might involve more layers of interaction. The problem is delineated in Fig. 6.1, which shows a conceptual view of selection and activation conditions.

The recognition region, folded away from the amino acid in the drawing, is shown in juxtaposition to the anticodon region of the tRNA. The major interaction may be between the loops of the RNA and the amino acids which, on account of their specific side-chains, may prevent or allow binding of another RNA and thus provide a selection mechanism. Only a properly binding RNA would bring the terminal alcohol group into proximity to form an ester catalyzed by an activating group ( a di-imide for example). The selecting feature could be the region that we now call codon and that would, depending upon the amino acid bound in the RNA loop, allow or prevent binding to the hypothetical transfer RNA. Presently, the cavity-forming RNA is replaced by aminoacyl-tRNA-synthetases which seem to occur in large numbers in the genomes of all organisms.[13]

Nothing described so far could persuade one to exempt the code-developing process from the common mass-action ways of chemistry. Consider, for example, that the code, which is collectively referred to as unique, actually consists of 20-some genetic codes, one or more for each amino acid. Twenty times at least did this series of complex reactions occur and all were alike in concept but specific in detail such as to distinguish between differences as small as between the side-chains of alanine and valine and as large as the difference between glycine and methionine or tryptophan. Fitting complex amino acid side-chains (methionine and tryoptophan) would be more demanding so that one should expect less ambiguity in their codons. The original

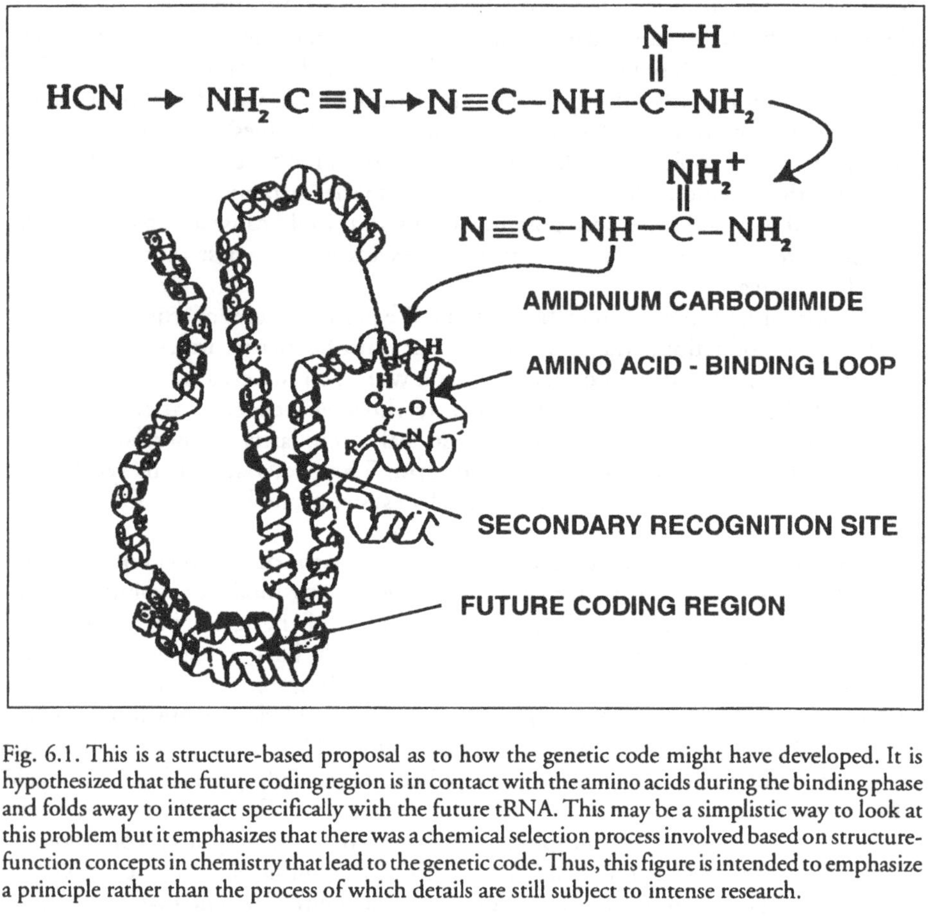


Fig. 6.1. This is a structure-based proposal as to how the genetic code might have developed. It is hypothesized that the future coding region is in contact with the amino acids during the binding phase and folds away to interact specifically with the future tRNA. This may be a simplistic way to look at this problem but it emphasizes that there was a chemical selection process involved based on structure-function concepts in chemistry that lead to the genetic code. Thus, this figure is intended to emphasize a principle rather than the process of which details are still subject to intense research.

RNA binding loop postulated in Figure 6.1 may have been specific only for L-configuration of the N-Ca-$C_{carbonyl}$ group, which is uniform in all protein amino acids (except glycine), whereas the RNA tail controlled the side-chain fit and thus the codon-specific interactions.

The code selection, a series of events that is cited as an irrefutable reason for a single origin for all species, happened at least 20 times! In the mind of a genomist, 20 repeats of a series of complex reactions, leading to the activation of 20 L-amino acids, means that there have been very strong determinants. Code development happened before cell formation was complete because cells need proteins. Could one imagine that one aggregate of RNA/DNA would develop a code for 20 amino acids while the neighboring clones produced none? Of course not, but plausibility is of limited help in a decision-making process in science. Beyond likelihood we must recognize that, save the reaction of oxygen and hydrogen, chemistry is not that sharply limited, certainly not in primordial bio-reactors. For distributions of products to be limited to one bio-reactor, the energy (entropy) of the system would have to have been prohibitive as compared to the surrounding; it would be at the level of supernatural!

But would all codes have to be alike in all developing units of life to produce a fauna and flora as observed today? Of course not, as long as a protein is made of

L-amino acids the feeding process would be undisturbed. Life as a global phenomenon would have functioned perfectly well if every species had its own code for producing proteins.

The actual code development must be viewed as a process, quite apart from, and independent of the accumulation of genomic material. The crucial event in code development is the direct or indirect interaction (Fig. 6.1) that ends with the covalent linkage of an amino acid with a very limited set of transfer molecules, all of which carry amino acid-specific triplets of ribonucleotides, called the anti-codons, at the tip of a hairpin turn. This is the beginning and the end of the codon selection process; it is all there is to it. The term anti-codon is not strictly correct because DNA (or the corresponding RNA transcripts) contains no codons. It is merely a string of deoxyribonucleotide phosphates whereas the genetic structure one refers to routinely is brought to life by the tRNA, which divides the string into three's, and by the ribosome, which stabilizes only one incoming tRNA at a time. Signals for start and stop of proteins are merely inconspicuous parts of a monotonous polymer until they are converted by recognition factors (feed-back proteins) to vital signals in a humming center of growth, control and maintenance of an organism. The sequence of the DNA determines the organism; tRNA and ribosomes validate and translate the information. Thus, a string of nucleotides such as AUUACACCGAACAAA reads nothing but when the codon adds punctuation, three at a time, i.e., AUU ACA CCG AAC AAA then the sequence reads Ile-Thr-Pro-Asn-Lys, which is a defined string of amino acids. A human genome read by the ribosomes, tRNA et cetera from bovidae would produce a human whereas a bovine genome read by the human translation machinery would produce a cow. That is what is meant when one refers to universality of basic constructs in life. It also vividly illustrates the high degree of independence of the coding process from the information carrier, the genome.

To explain the L-amino acid preponderance, which enters into the code production, processes such as parity violation at the level of electro-weak forces[14] have been considered. That proposal is likely not correct because the signal is too weak and because selectivity based upon atomic properties seems to be important, rather than quantities. Carbon was not the most prominent element on earth by far when life assembled itself from carbon rather then silicon.

The role of the initial binding RNA in Figure 6.1 is now fulfilled by aminoacyl-synthetases which were selected from a number of proteins of suitable activity. An inordinately large number of this type of structural motif is found in different genes which supports the basic concept of function selection (13) as opposed to the development of function by targeted mutations.

The genomic DNA is of necessity without design. All of the information contained within the strand in form of sequence variability comes to light through the code. One could state that the nearly limitless potential information hidden in the tons of nucleic acid of the DNA or RNA type, regardless of which code would be adapted, would inevitably lead to all the life forms on earth. The total variety of life forms on a particular planet under proper biogenesis conditions becomes a function of the total amount and variability of nucleic acids available. That gives us a bio-potential theorem worthy of the new millennium.

Whatever the outcome, we can rest assured that a physical contact type selection (chemistry) has led to the genomic code because it is, with insignificant variations, the only code possible. Life anywhere in this universe will be C, H, N, O-based, and the genetic code will be found to be the same as well. If research would show that there is no structure-based connection between the genetic code and the structure of the amino acids and nucleic acids involved, then the multiple origin idea is, in my opinion, untenable and evolution would not be subject to scientific pursuit.

The march toward the world of proteins is another complex story of timing and contact. And again, well-behaved energy was needed, a gentle stream of useful quanta that would not destroy the molecules through which they were captured. The chemical bond, particularly in conjugated systems, can be activated by visible or UV light to provide excited electrons. The electrons need to be passed along to receivers that can convey them to reactants such as to push a nonproductive equilibrium into producing a steady supply of building blocks for biological structures. This passing along of electrons is usually done by proteins and supported by cofactors such as complex conjugated ring systems. The chain of electron transporters needed to be assembled under prelife conditions such as to kick in at that critical moment of transformation to life. Our knowledge of biology pushes us on and on to more assumptions. Where is the bottom of this story?

The early amino acid condensations must have occurred without enzymes, or rather without proteinaceous enzymes, because the proteins are the products of that process. Proteins are not very stable molecules and need to be reproduced continuously to keep a certain catalytic process going. They are 'virtual' components of living systems that exist only as long as their coding sequence is active. The relative stability and accurate reproducibility of the nucleic acids that would eventually become the genome would guarantee the continuous production of proteins under biological as well as prebiological conditions. Nucleic acids were the original template from which everything else derives. They are the reason for the variety of life forms, the uniformity of basic functions, and the stability of species.

It follows that the coding machinery must have come into existence in the prebiotic world in order to produce the catalytic peptides to support minimal metabolic activity and communication at the moment of cell formation. Many of the prebiotic associations of molecules might seem fortuitous when more likely the high redundancy of nucleic acids offered a great variety of proteins produced without a target. Eventually metabolic pathways will develop from the large offering of nucleic acid messages (proteins).

The Embden-Mayerhof pathway of glycolysis, also known as anaerobic metabolism, is perhaps an early function coupled directly to photosynthesis, which requires chloroplasts for the recovery of reduction potential from the photolysis of water.[15] Chloroplasts were perhaps finished even earlier than any other structure but certainly 3.5 billion years ago lest blue-green algae could not have been blue- green!

As concerns communication functions, the first proteins had to be ready for their roles as osmotic and nutrient traffic regulators at the moment of cell closure. This again is most plausibly achieved by uniting smaller open catalytic units, referred to above, to form a closed functional body. Each of these units would contain genomic material to produce the proteins associated with it, and that would be the contribution to the start-up package for the new cell, the first potluck event on earth, as it were. The diffusion distances are very short in concentrated solutions so that a prebiotic commune survived by sharing even before they united to form cells. Micro-organisms have a one-compartment structure, precisely what one would expect to get first when one would roll into a unit some of the catalytic foci that were floating in a concentrated suspension. The cell has all the making of an aggregate with interwoven lipid layers which, because of their chemical properties in water, tend to be double membranes as we see them in chloroplasts and mitochondria.

Once the early code reader had been produced from RNA, which is the active component even in modern ribosomes, the potential information stored in the nucleic acid chains became defined and accessible and the protein products could be recruited as expeditors of the expression of a limitless reservoir of information. The reiteration of complementary strands of nucleic acids never stopped even when all possibilities of every possible protein plus all the failures were produced many times. All structural motifs were exhausted and all of them were potentially available to nascent cells; they just had to be there to be collected in a grab-bag fashion. How much nucleic acid material was there? Was it enough to buy all the tickets in the lottery?

The amounts of material of almost any description falling onto the earth crust per year is astounding. Between 500 million to one billion tons of nitrogen reach the earth per year. If the abiotic synthetic period produced only 10 tons of nucleotides that would be a modest estimate. One mole of nucleotides weighs (rounded off for simplicity) approximately 200 gram and the 10 tons would amount to 1000 moles or $10^{26}$ molecules which, made into a string (10 Å per nucleotide) would cross a substantial portion of our galaxy. Stepping along this string, codon by codon, would give one a fair idea about infinity, about the unlimited information available at the origins of life.

Images are almost as important as any argument when it comes to extreme stories. The best of presentations would not be understood if the audience cannot imagine the new concepts in worldly pictures. Can one see one's own genome on the tip of a needle with all that empty space around it? This picture introduces us kilo-sized organisms to the enormous space that exists in the submicroscopic world and this is the world where biogenesis occurred. It says that within the expanse of a needle tip plans for many creatures could exist so that mixing and matching among pregenomic nucleic acids is well within the range of diffusion. The linear information would have to have been available in a small space in order to provide the benefit of limitless information to small units like the primordial cell Anlage in statu nascendi. The size of the human genome is about $3 \times 10^9$ base pairs and the length would be $3 \times 10^9$ times $10^{-10}$ meters, which amounts to $3 \times 10^{-1}$ meters or 30 centimeters. Of the $3 \times 10^9$ base pairs, only $3 \times 10^7$ (30 million or 1%) are needed to spell out a human being! That would be about 3 millimeter of DNA. Mixing and accessibility during construction of a gene depends upon the bulk as well as the linear dimension. The 30 cm of the human genome is so narrow as to be invisible unless viewed through an electron microscope.

Cells would have persisted only because of extreme efficiency in self-maintenance. In cell compartments the catalyst concentration would be a hundred to a thousand times higher than on the outside where everything is large compared to the picoliter size cell. A few protons, two to three in a lysosome (a digestive vacuole), would produce pH values of 1 (the equivalent of 0.1 molar hydrochloric acid), and two enzymes in that space would bring its concentration into the millimolar range. In a cell the affinity of the enzyme for its substrate determines the speed of the reaction as opposed to thermodynamic limitations of the chemical transformation *per se*.

It is very plausible that all of life should start with blue-green algae which were the only creatures around that were able to concentrate the diffused solar energy sufficiently to support growth and development. According to currently prevailing ideas, these organelles (pro-chloroblasts, pro-mitochondria) crossed the threshold to life as independent prokaryotic units and the residual nuclear material in them resembles the modern prokaryotic gene rather than the eukaryotic one. The mitochondria, the mediators of oxidative phosphorylation, are problematic for the Darwinians, not because they formed but because they formed when, according to their model, there was no "evolutionary pressure"(oxygen) to produce them. Oxidative centers of complex biology have developed when no oxygen was around which means that their assembly was not stimulated by adaptation to oxygen but was rather the consequence of a certain nucleic acid configuration, the product of un-erasable redundancy. Possessing these oxidative enzymes during the anoxic period was not a severe flaw, but became a tremendous advantage when oxygen levels began to rise. Not only could these organisms deal with oxygen toxicity and corrosiveness, they were able to oxidize hydrogen in a controlled fashion making off with 57 kilocalories of energy per mole parceled out in small and biologically acceptable units for the synthetic activities.

There is another line of threshold activity required, namely the fixation of nitrogen. Indeed, the blue-green algae that were just considered a single cellular organisms may have shown a subtle beginning of cell specialization. Every eighth unit in a chain of blue-greens is a heterocyst that fixes nitrogen, neatly separated from the incompatible oxygen-producing capacity. This again is an activity that could not have been anticipated, it was the innocent product of aimless redundancy sorted out by compatibility.

Combination and endless reiteration of early messages is the concept that takes the mystery out of biogenesis. All basic functions are very similar as chemistry would dictate, but the subtlety of the same chemistry expressed in nucleic acids provides for different organization of the genome and thus causes different species to appear. No plan, everything that works together well will persist until the condition changes beyond the limits of flexibility of a unit at which time extinction becomes the only route; the fossil record agrees.

Further development comes from the organization of nuclear material into more efficiently functioning units. All appear as prokaryotes at first but after several hundred million years of rummaging through the genomic material some began to organize the genomic material around positively charged proteins that were encoded in some of the sequences uncovered by internal reorganization. Thereby much larger genomes became accessible for protein-read off and thus some cells were able to increase their sphere of living space by adding a balcony of protoplasm around the nuclear material. In the process genes that encode lamin and other nuclear envelope associated proteins were uncovered and activated by some eukaryotes and that seems to have been a key-step towards multicellularity.[16] The age of eukaryotes, the animal and plant cell type had begun.[17] In fact, reports are presently appearing that push back the range of recognition of eukaryotes to more than 2.5 billion years; their Anlage, however, dates back to the biogenic period about 3.5 billion years ago.

A mechanism of biogenesis must account for the fact that life is instantaneously demanding and there is no time to produce any vital component once the membrane has closed around a nascent cell. Each unit is sparked into being in seconds and if a burst of light were associated with that instant, a long-lived observer walking across the early earth would see for a few million years these flashes dotting the landscape.

He would not be aware of the thousands of preparatory chemical steps that precede each finished cell. When scintillation stopped, the biogenic period was over for this planet. The bio-earth had created itself and every life form had started on its billion-year trek to fulfil its potential as conditions changed.

## References


1. Mayr E. Darwin's influence on modern thought. Sci Am 2000:79.
2. Benton MJ. Diversification and extinction in the history of life. Science 1995; 268:52.
3. Nursall RR. On the origins of the major groups of animals. Evolution 1962; 16:118.
4. Keosian J. Life' beginning—Origins or evolution? Orig Life 1974 ;5:285.
5. Kerkut GA. Implications of Evolution. London and New York: Pergamon Press, 1960.
6. Berg RL. Some conditions for the appearance of life on earth. In: Clarke F, Synge RLM, eds. Proc I Internat Symp: The Origin of Life on Earth. New York: Pergamon Press, 1959:169.
7. Schopf JW. Microfossils of the early archean apex chert: New evidence of the antiquity of life. Science 1993; 260:640.
8. Cech TR. Ribozyme self-replication? Nature 1989; 339:507.
9. Hendry LB, Bransome ED, Hutson MS et al. A newly discovered stereochemical logic in the structure of DNA suggests that the genetic code is inevitable. Persp Biolog Med 1984; 27:623.
10. Woese CR, Dugre DH, Dugre SA et al. On the fundamental nature and evolution of the genetic code. Cold Spring Harbor Symposia on Quantitative Biology 1966; 31:723.
11. Welton MGE, Pelc SR. Specificity of the stereochemical relationship between ribonucleic acid-triplets and amino-acids. Nature 1966; 209:870.
12. Pelc SR, Welton MGE. Stereochemical relationship between coding triplets and amino-acids. Nature 1966; 209:868.
13. Schimmel P, de Pouplana LR. Footprints of aminoacyl-tRNA synthetase are everywhere. TIBS 2000; 25:207.
14. Avetisov VA, Goldnaskii VI, Kuz'min VV. Handednes, origin of life and evolution. Physics Today 1991:33.
15. Margulis L. On the origin of mitosing cells. J NIH Res 1993; 5:65.
16. Cohen M, Lee KK, Wilson KL et al. Transcriptional repression, apoptosis, human disease and the functional evolution of the nuclear lamina. TIBS 2001; 26:41.
17. Brocks JJ, Logan Ga, Buick R et al. Archean molecular fossils and the early rise of eukaryotes. Science 1999:1033.

# CHAPTER 7

# The Origins of Species

There was never a time on earth when only one kind, one species, existed. At least there is no evidence to that effect and a plausible extension of that simple thought would lead to the conclusion that the development of species was parallel rather than sequential. The old paradigm starts with a single origin and in order to bring nature into harmony with that assumption, they invented speciation, to designate the splitting of species into new, biologically isolated units. Obvious as species are, speciation has never been substantiated and all examples given in the literature are postulates inspired by the hypothesis. The paradigm says that if animals A and B have certain features in common they are derived from a common ancestor. Thus, the evidence available is merely a restatement of the parent hypothesis.

During a conference on evolution in Prague in the '70s when a session on speciation began to grind to a halt I was asked to resuscitate the fighting spirit of the group. I do not remember why the onus fell on me after the organizers had allowed just minimal exposure for my thoughts on the program. There I was facing this group, their eyelids half folded, limbs hanging from the arm rests when I heard my own voice suddenly and unconnectedly asking whether any one in this group had ever seen unequivocal evidence for speciation. In a session on speciation this was the stone of Jason. Several jumped up and gave examples of multiple forms of a species that were tautologies on close examination, i.e., the hypothesis says that two similar looking animals came from one stem and therefore there was a speciation event; nothing really made a point. Finally one member of the group attempted a coup de grace for my contentiousness. Two colonies of fruit flies, he said, had been bred apart for the equivalent of millions of human propagation cycles and when brought together they would no longer interbreed! This was the core of the Darwinian idea confirmed in a simple indisputable experiment. What was there to say? Both groups still looked like *Drosophila melanogaster*, true, but it might have been seen as the initial stage of speciation; there was nothing left but to concede. At that moment a voice from the ranks broke the momentary silence, announcing that these experiments had long been shown to have been flawed and yes, the two groups were interbreeding as if they had never been separated. The session ran overtime without producing a single example of speciation. The excitement continued into the dining hall, but the lesson was lost over excellent Pilsner.

Yet, species are real and I owe the reader an explanation as to how they might have come about in the new model, particularly since earlier I argued for uniformity in chemistry. And by what means would one find an explanation if not by the same fossils that Darwinians are using combined with a new philosophy.

Even in our sort of orthodox climate of evolutionary thoughts, experience teaches that fossils that are never unequivocal on nearly anything are very clear in denying the Darwinian process of speciation through imperceptible changes. Quite intelligent people are overlooking this conclusion because the evidence for the Darwinian model will be discovered, they are made to believe, during the next field trip. Of course that is an irrefutable argument, but it is also a worthless one. After 150 years of fruitless search it is prudent to conclude that the intermediates of the kind postulated by Darwinians do not exist.

Before even looking at fossils the genomist notes that evolution of animals via intermediates would be incompatible per se with the existence of species. The "inconceivably" large number of intermediates[1] that anchor the old theory would have eliminated all species distinction and produced a bio-continuum instead, which is not observed. Furthermore, the fossil record shows species retaining their identity for millions of years, and that observation certainly does not support the concept of fluid barriers between them. Peculiar as it sounds, the Darwinian hypothesis in its original form is incompatible with the existence of species. On the other hand, species defined as propagating units are the only natural division of the fauna and flora; all higher-order taxonomic divisions are a way of grouping species by similarity, which may have no biological meaning

Species became apparent in the Ediacaren, but the picture is even more persuasive in the middle Cambrian period where the first arthropods were appearing, dense and varied, within a few centimeters of the lower layers of the Burgess shale.[2] There is a splendid display in the Smithsonian Museum of Natural History of innumerable animals that appeared in a very short time, but the consequence of this observation has not even penetrated to the core of paleontologists' paradigm which calls for about 10 million years for each speciation event. The reader may imagine a miniscule arthropod which would have been the ancestor of all others and therefore the beginning of the metazoan era. We will ignore the question as to how it propagated and come back to this spot in the shallow Cambrian waters about 10 million years later. On the Darwinian timetable there would now be two different kinds of arthropods, the one seen before, which had propagated into larger forms, and the smaller newcomer. This would have been the first speciation and, as one now visits the Cambrian every 10 million years it is noticed that, by the time the first road signs to the Ordovician appear, only six speciation events (10 million years each) have happened, which means 32 species materialized during the Cambrian as shown in Fig. 7.1

This conjecture collides with 30 major forms and 120 genera found in the bottom layer of the Burgess shale, which were deposited within the time span of a few million years. There they lie on top of each other like a cake of compressed creatures of the past.[2] Adding to it the corresponding layer of the Cambrian Chengjiang fauna of China[3] one cannot help but suspect that speciation had happened already long before the Cambrian or Ediacaren, although there are no hard fossils to confirm that suspicion. I have made a point of this message to my audiences and have noticed that it registered here and there.[4-6]

Of course, there are the calcareous microfossils of the Tommotian[7] and the Precambrian,[8] but nothing that would be an obvious immediate set of ancestors to the Cambrian assemblages. One knows that antecedents are there because of the laws of cause and effect, and we have even seen them in their premetamorphic phase without recognizing them because until now there was no reason to look for any such stages of life.

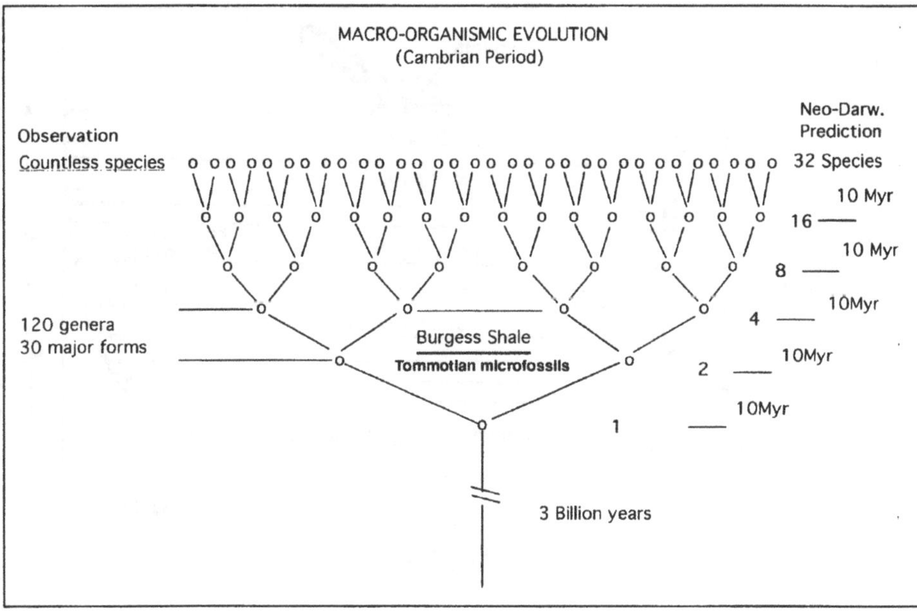


Fig. 7.1. The old model is driven by the idea of "descent with variation," meaning that all organisms start from one event, followed by speciation about every 10,000,000 years. For those of us who do not trust mathematics, this scheme shows that the process would produce 32 species from the beginning of the Cambrian to the end of that period, assuming that no extinction had occurred. This observation has to square with a reality that produces 120 genera of 30 major forms in the immediate post-Tommotian segment of the Cambrian Period.

The Genomic Potential Hypothesis should inspire paleontologists to search for instances of contiguous discontinuity (like in caterpillar and butterfly) in the layer that underlies the first fossiliferous horizon. This is a tall order, but recently the embryos of worms[9] and possibly of arthropods[10, 11] have been discovered in the upper Precambrian layers. Micro-structures, recovered from deep phosphate-rich deposits in China and Siberia[12] appear just like the pro-forms of animals might have looked that antedated the outpouring of phenotypes in the Ediacaran and the Cambrian which were followed periodically by more or less intense displays up to the late Quaternary when upright walkers appeared. This would suggest that the Cambrian animals did not come from the Ediacaren group but derived more likely from germinal forms similar to those that had given rise to the Ediacaren fauna except that the Cambrian pro-forms were a little more complex! The pro-forms for the Devonian creatures were even more complex and remained silent for another 200 million years. My readers can continue the idea now until the Neogen when their own pro-form began to break into phenotypes. And why would that be a more reasonable conjecture than speciation by branching in the old model? Because there is evidence for the pro-form/phenotype transition whereas there is none for branching!

In proper perspective, i.e., in relation to a 3-billion-year-long developmental period, the distance on the time scale between the eruption of trilobites versus hominids is remarkably short. It is much too short to build a human from an arthropod by gene duplication and mutation!

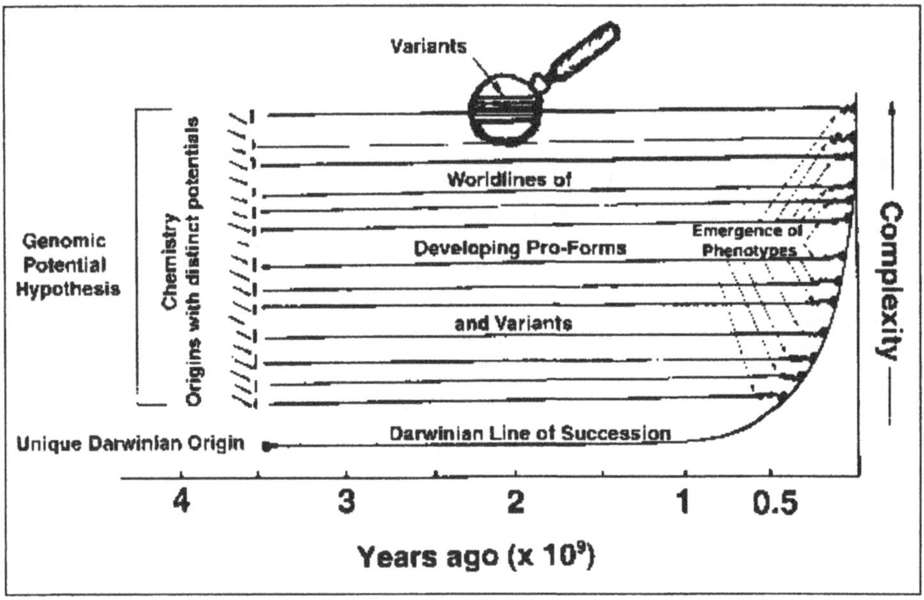


Fig. 7.2. This figure shows schematically the major difference between Darwinism and genomism. According to the new hypothesis, every life form starts its march to the present during the one biogenic period on earth, which occurred 3.5 billion years ago. The length of the time of development is determined by the complexity of the Anlage. The least complex are microorganisms which are essentially finished when the first stage of life appears in the Hadean rocks. The simplest macro-organisms appeared during the Ediacaren, followed in short order (by geological standards) by increasingly complex forms during the Cambrian and all periods thereafter. Each group has its own hypothetical metamorphic period. The contrast with the pluripotent origin of the old model is striking in many aspects, most remarkably when one considers the very short time period of development allotted between successive layers of complexity in the Darwinian model.

Figure 7.2 depicts the Darwinian model as a single line from a single origin that swings up sharply at the beginning of the Cambrian. The multiple origin lines of the Genomic Potential Hypothesis symbolize genomic development that lasted 3 billion years for each life form, ending in the metamorphic state (wavy line) and finally in the same creatures that the Darwinian model must develop, one from another, in the phenomenally short period of 500 million years. The phenotypes of the Quaternary (our period) are so different from the Cambrian creatures, it is argued by Darwinists, that only a long time of adaptation, fine-tuning so to speak, by natural selection would have converted them into all subsequent species. One must object to that proposition on the following grounds: As the Precambrian stem cells burst onto the animal scene they brought along from the single-cellular (or colony) stage their appendices, sensory and reproductive organs, their feeding machinery, and armor for protection. But since these features, save reproduction, are without meaning in single-cellular life from which the animals just emerged, their appendages could not have been adaptations to any need for defense that multi-cellular life forms developed purportedly to deal with each other. What kind of evolutionary pressure would have prepared single-cellular organisms for what lay one short step ahead of them once they passed the line to multicellularity? The well-articulated legs and claws were an expression of genomic configurations that produced chitenous phenotypes directly and in prodigious amounts in the Cambrian. It was the chitin age; bone was a little down the road. Naturally one would ask if claws can come about at this stage on the basis of molecular development rather than adaptation, could not an arm come about by the same route a little later?

During Darwin's time, St. George Mivart called attention to the difficulty entailed in the concept of adaptation by selection but it took 150 years until the modern evolutionists noticed and then they merely invented another term called preaptation instead of tossing the hypothesis.[13]

Why should one consider adaptation as an explanation for the production of new forms when there is no molecular mechanism and no material evidence? Extinction is prominent in the fossil record but not adaptation[14] which is a concept that is a nonlinear extrapolation of physiological adaptation, such as the increased myoglobin levels observed in the muscles of diving mammals, hemoglobin in mountain tribes, and muscle mass in athletes. These permissible adaptations always involve an increase or decrease of what is already part of an organism. The development of new body-parts is an entirely different matter that would, were it possible, involve changes in homeotic genes that control segment activity and coordination.

Metamorphosis, the process whereby one creature comes from one genome in two or three distinct forms called egg, caterpillar, and butterfly, is a well-recognized process because it can be observed during every phase. It is a prime example for phenotypical development by molecular design without guidance from Darwinian principles.

The transition from a primary ovum to a phenotype is an inescapable conclusion when all circumstances are considered. Just as clear is the fact that between the DNA sequences and the carapace of the first arthropods bilateral and multilateral symmetry fell into place without help from the environment.[15] Most animals are symmetric, but there are enough exceptions to make the point that this is not due to 'evolutionary pressure' but is rather a consequence of a molecular configuration at the genomic level.

So what is the gist of all these explanations? They serve to make room for a new interpretation of the fossil record. According to the new model speciation was a fait accompli when the first animals began to produce fossils. It happened even earlier, the true edge-condition for speciation must be placed at the Hadean-Precambrian transition and that means that it was a chemical and not a biological phenomenon.

There are no species (as in life) before the Hadean transition, but the root of the argument concerning the segregation of future life forms (species) actually begins when nuclear raw material formed nucleotide sequences which would determine the possible outcomes of the developmental period of a cell. This is the absolute bottom of the story, the time when the genomic potential develops. Before polymerization, nucleotides do not carry any information, once they are linked into chains they spell out an organism if exposed to the proper conditions, and they remember their own sequence because of the complementary type of self-replication as plus and minus strands. There is a true border condition between monomers that carry information concerning chemical reactivity only and the polymers that hold the information for a macro-organism. For each future clone that consists of thousands of nearly alike nucleotide sequences in an area of biogenesis we can write the equation of species:

(Nucleotides 1, 2, 3,—n)—polynucleotides—potential species + variants and so on for a vast number of biogenic foci.

The history of species begins in pools which were likely not much larger then a liter and which might have seen periods of concentration and dilution and what else the relatively rough climate of early earth presented. At the nucleotide level there was no readily discernible difference between the pools but when polymerization began every pool was different from its neighbor and chains of different sequences arose even in the same pool. When primary chains reached a certain length hydrogen bonding between a polynucleotide strand and free nucleotides would promote synthesis of complementary strands which would give a certain identity to a focal point. Information potential would increase with increasing length of the primary strand much like the information expressed as 0 and 1 on a Turing tape would increase with length of the tape. All of the processing of potential information would occur by the reactions that are well-known to molecular biologists such as loop formation, scission, fusion, and insertions. Furthermore, assembly of functions from a huge reservoir of nucleic acid, produced by untargeted polymerization, requires that large amounts of useless sequences exist. From what is known about the human genome that means 99% of the DNA in the nucleus.

Another interesting conclusion would be that such an assembly of untargeted information would contain a great deal of redundancy of important and of unimportant proteins. There are about 40 fibroblast growth factor genes of which the major forms can be experimentally eliminated (gene knockout) without any ill effects on the organism. A major blood protein in humans (albumin) is not required for survival or well-being, and relaxin appears in humans in two copies both of which are not essential. The pan-selection concept of the old model does not permit the production of useless products and modern gene technology proves them wrong almost daily by showing up unessential proteins. Clearly the organism is not constructed by reason and the remaining possibility is the assembly via an excess of potential functions. That observation leads us back to the prebiotic origin of the only natural divisions of the fauna and flora, the species, and there is really no viable alternative, nothing that could stand up to the demands of epistemology. This line of reasoning leads to a universal concept of biogenesis and bio-diversity, which states that the complexity and diversity of life on a planet is a function of the amount of nucleic acids that have accumulated via chemical synthesis.

## References

1. Darwin C. The Origin of Species by Means of Natural Selection. New York: The Modern Library of New York, 1859.
2. Morris SC. The Cambrian "explosion": Slow-fuse or megatonnage? Proc Natl Acad Sci USA 2000; 97:4426.
3. Chen J-Y, Ramskoeld L, Zhou G-q. Evidence for monophyly and arthropod affinity of Cambrian giant predators. Science 1994; 264:1303.
4. Ohno S. The notion of the Cambrian pananimalia genome. Proc Natl Acad Sci USA 1996; 93:8475.
5. Doolittle WF. Phylogenetic classification and the universal tree. Science 1999; 284:2124.
6. Woese C. The universal ancestor. Proc Natl Acad Sci USA 1998; 95:6854.
7. Stanley SM. Earth and Life Through Time. New York: Viking Press, 1986.
8. Schopf JW. Solution to Darwin's dilemma: Discovery of the missing Precambrian record of life. Proc Natl Acad Sci USA 2000; 97:6947.
9. Chen J-Y, Oliveri P, Li C-W et al. Precambrian animal diversity: Putative phosphatized embryos from Doushantuo Formation of China. Proc Natl Acad Sci USA 2000; 97:4457.

10. Pratt BR, Zhang X-G. Unscrambling fossilized Cambrian embryos. J NIH Res 1995; 7:46.
11. Bengtson S, Zhao Y. Fossilized Metazoan embryos from the Earliest Cambrian. Science 1997; 277:1645.
12. Li C-W, Chen Y-J, Hua T-E. Precambrian sponges with cellular structures. Science 272; 279:879.
13. Lewin R. Adaptation can be a problem for evolutionists. Science 1982; 216:1212.
14. Raup DM. Biological extinction in earth history. Science 1986; 231:1528.
15. Peterson KJ, Davidson EH. Regulatory evolution and the origin of the bilaterians. Proc Natl Acad Sci USA 2000; 97:4430.

# Chapter 8

# Development of Biological Potential

Should development come after speciation? Yes, of course, in the new model it is natural, one produces the Anlage and develops it to its potential. The word evolution is derived from the latin term *evolvere* which means to roll out and that describes precisely the process postulated in the Genomic Potential Hypothesis for the post-Cambrian time. With speciation behind us there remain two distinct phases of evolution in the widest sense of the word, the streamlining of the chemistry in the nucleus, which leaves no traces other than species-specific stem cells, and the post-Cambrian 'unrolling' of species and variants that produces a spectacular display of phenotypes. Looks like we are on the right track.

About 200 years ago it was clear that life's history began in the Cambrian and whatever was before, long or short, was a blank called Precambrian. Even in the 1994 a nonreligious book appeared, the author of which, de facto placed the origin of life into the Cambrian.[1] Darwin sensed the need for more time than 500 million years for his evolution model and that conflict led to an argument with Kelvin who was unaware of radioactive decay and its heating effect upon the earth and had, as a consequence, miscalculated the age of the universe to be no more than 500 million years. The naturalists at that time would have missed the 3 billion-year period of evolution anyway because the methods to find fossils of micro-organisms and to determine the age of the rock matrix had not been developed.[2] What difference would it make to know about a period during which nothing overtly happened? It provides a credible basis for a scientific hypothesis of evolution!

Although the first cells and the first macro-organisms are separated by an unimaginable stretch of time, without these early events the planet would have been arid today. Archean rocks, however, reveal an astounding assembly of micro-organisms,[3] many of them sufficiently familiar to find a place in today's systematic taxonomy. Life flooded the underground scene to begin the longest bio-period known to us. The old model retrofits this stage by calling it a stasis of 3 billion years. True, the fossil record shows 3.5 billion-year-old cyanobacteria, looking like those 800 million years of age as well as contemporary ones, but that does not really mean that nothing happened, it only shows that cyanobacteria did not change.[4] This pattern continues throughout all of the evolutionary history; as soon as a species is recognized in the fossil record it will no longer change significantly. Most macro-organisms were not finished until the Cambrian but then they shattered the tranquility with an avalanche of animals strewn over the shallow Cambrian seas around America, Australia, and all the way to what is now the eastern part of China (all in the Pangea[5] configuration). Prokaryotes remained prokaryotes and today they constitute perhaps the most abundant and most adaptable form of life. They are reproductively so successful that any development away from prokaryotic life must be considered a significant step backwards in terms of survival capacity. Potential eukaryotes were invisible for us during the first billion years of the Hadean. The potential to produce, either protozoan, plants, or animals is recognized only when it is expressed.

If indeed the potential for all organisms was there at the moment life appeared, what took them so long to get up and running? Energy minimization is a possible answer, but to make sense of it I must appeal to the power of imagination which lets one penetrate the cell and perceive its minuscule dimension as a cavernous space with chemical reactions occurring in the polymer phase at many different places at once. Looking through a nuclear pore of a eukaryote that lived 2.5 billion years ago[6] one would see long strands of DNA and RNA turning and forming loops which occasionally break off, only to be reorganized and incorporated into a different section. The field of view changes constantly; bundles of loops, hairpins and loose ends of DNA hopelessly entwined, jerking and twisting from thermal motion until an RNA loop drifts by, interacts with the bundled DNA for a millisecond upon which the knot falls apart, relieved of internal strain. Certain sections of the DNA have served as a template for new strands, but now some of the previously inaccessible loop sections that have been freed by internal catalysis will also duplicate. In this way one could imagine additional potential of a genomic configuration in a particular cell to be recovered for the production of catalytic or structural proteins. The reorganization of the DNA is a sequence-dependent phenomenon which, by the nature of this molecule, will cause loop-outs, self-cutting and back-splicing of cut-outs into another position until an equilibrium is established wherein forth and back reactions occur equally fast and the nucleus, or rather the DNA, will retain a particular configuration for a majority of its lifetime. Reorganization under the direction of energy parsimony provides the major orienting force, but the equilibrium changes as proteins are produced that catalytically expedite the process. Within cells from the same region these reactions will happen with the precision of a democratically run ballet, nearly the same way in every cell sooner or later. A different initial configuration of clones will lead to different final products and so on, for billions of surviving foci.[7] By this process, one might imagine that the basis accumulated for the expression of potential species and variations.

Imagine two long stretches of DNA which are alike except for a few positions in the center. It is energetically favorable for configuration A to bend back upon itself, reanneal its terminal ends to the middle of the molecule, cause a breakage, and then reinsert the broken piece in the middle of the loop. Molecule B cannot do so because the two or three different bases prevent annealing in the same position, but it can form a loop much smaller at the opposite end where it causes a scission and reinsertion.[8] When we compare the linear sequence of these hypothetical DNA polymers, the original minor differences have now been increased substantially because of one step of reorganization that was predetermined by the stability of a configuration but could not be anticipated from mere inspection of the DNA pieces. The process, if carried out over eons, would cause molecules that were similar at the start to be significantly different at the end of the 3 billion-year reorganizational period without the need for a single mutation. Again it is important to remember that no goal is being pursued; the rearrangement leads the nucleus to a less energetic conformation. These processes are restricted to the cell, nothing escapes and nothing enters; the genomic material merely reorganizes according to its potential in the direction of the lowest free energy. Pro-eukaryotes did evolve at the same time as prokaryotes but had a different developmental potential, which became obvious when the oxygen levels in the atmosphere increased significantly. In any event, subsequent eons belonged to genomic refinement, i.e., evolution in the spirit of the Genomic Potential Hypothesis.

And what may drive refinement in this world that knows of no encouraging canines or poisons, summarily called evolutionary pressure, to force progress?

The products of efficient reactions will accumulate faster than the products of inefficient ones and therefore the most robust processes will dominate, no urging required. This is how efficient life forms came about at the cellular level, and this is also the reason why 3 billion years later well-functioning and efficient forms of macroscopic life (trilobites lived more than 50 million years) appeared suddenly,[9, 10] peppering the Cambrian landscape.

The selection is based upon favorable reactions; transcripts may reenforce their own messages and thus establish a hierarchy that forces development into a particular direction so that the low energy-state now includes the new proteinaceous catalyst. This process took 3 billion years, which is understandable considering the complexity of a nucleus.

Chemistry is a very fast process and one might wonder why the rearrangement should have taken all of 3 billion years. Analogies provide no answers but they reinforce one's imagination, and in this case "Rubic's Cube" might help to explain why genomic streamlining took so long. Starting with a cube in its most disordered state one begins to twist and turn until the most ordered state has been reached (all red surfaces out, for example). The energy we spend depends upon the length of time it takes to reach that defined end point. So it is for chemical or physical processes which will always tend toward low energy positions, provided that a kinetic path exists between the two states. But the path can be full of minor energy minima, and the process could therefore get stuck in intermediate positions for very long times. The cube was designed to have a path from state A to B, which the initiated will complete in a few minutes. But for the uninitiated the path may be extremely long or never-ending because of many wrong moves that have to be corrected by backing off and starting over again.

Symmetry in biology is one of the many unresolved problems, but complex examples can be observed in the inanimate world. As dust specs are drifting through the wintry sky, water molecules freeze to the surface to form a delicate crystalline marvel of precisely sixfold symmetry.[11] Deterministic? No doubt! The architecture of each of the six identical leaflets in one flake is determined in part by the nucleating surface and by the temperature gradients through which it tumbles. While it is said that no two snowflakes are alike, the sixfold symmetry is invariant.

The assembly of animals contains an element of lateral symmetry as well as longitudinal orientation which may be represented by a string of letters that designate complexity factors and subscripts that are symmetry factors whereby 2 means bilateral—and 3,4,5 —n multi-dimensional symmetry. Thus, a bacterium is $A_1$, a starfish $A_{(5\text{-}9)}$, and a bony fish might be $A_2\,B_2C_2D_2$ + SCF, (Segment Control Factor).

The reader may blend in the scenario of nuclear reactions and be reminded that this is where biological symmetry originates. Let us assign completely imaginary complexity labels, A, B, C, D, E ——————N, to a potential organism. Level $A_1$

is the starting position most frequently attained and that would be typical in microorganisms which have no fixed major axes of symmetry. One could assume that symmetry factors terminate the complexity chain growth and that, if an A could associate with a B before a symmetry factor would prevent lengthening of the string of letters, then a more complex organism could result. Thus A, B, and C would have to find symmetry factors which would take more time than for A and B alone and so on. The original genomic organizational level might be $A_2$ $B_2$ $C_2$ $F_2$ $G_2$ H I J $K_2$ L M N O P Q $R_2$ S, which would be the foundation for a complex 'bilateral', provided that the second unit is found for those letters that do not have a subscript. In contrast $A_2$ B $C_2$ is the basis for a simple organism. Required for phenotypical expression is an unbroken series of complexity and symmetry factors and, clearly, it would have taken more time to complete the long series than the small one; such a model would cause simple organisms to appear first and complex ones last. Note that this proposal matches observations and explains in concept why the human fossil record is about 160 million years shorter than that of a frog. It may be a simple model but it addresses a functional aspect of organogenesis and is therefore quite advanced compared to the standard model that advocates searching for Lucy's (*A. afarensis)* grandmother in a litter of tree shrews.

Energy minimization of the genetic material in the nucleus can be displayed as an asymptotic curve that edges ever closer to a minimum. Upon close scrutiny innumerable small energy valleys may be seen superimposed on the general trend wherein a conformation could get stuck for a long time. This roughness is related to the tremendous size of a genome. If a favorable conformation occurs on one end and another one at the other end, upon coming together those two folds may turn out to be unfavorable for the overall energy. Hence the large nucleic acid molecule can go through many conformational and covalent bond changes that, just like turning sections of Rubric's Cube, would have nearly equal energy differences, and a wrong move at one point might similarly require a return to another starting configuration before successful reorganization can occur. For eukaryotes the final state must be so efficient that the 3 billion base pairs of the genome can be unrolled, duplicated, and reorganized in the few minutes of a mitotic cycle. For that feat basic proteins had to be recruited about which nucleic acid could be wound and unwound in the few minutes of a duplication cycle.

Complexity increases if one factors-in the effects of the gene products, namely the proteins that form cellular structures or biological catalysts. Some of these proteins do regulate the expression of genomic material and in the process combine with DNA to either inhibit or promote the translation of the DNA into proteins. Thus, the gene products enter this equilibrium toward conformational and structural stability and the speed with which these reactions occur may cause the genome to stay on one of the conditional energy recesses where it is more responsive to control than it would be in an absolute minimum state. There is no evidence for this suggestion and our level of experimental sophistication is not sufficient to produce any. In principle, this problem can be addressed because it is embodied in a structure.

Arrival at this metastable state wherein the existence of a genomic configuration depends on the continuous rapid production of proteins was the signal, the moment when macro-organisms might have arisen from cells. The simplest nuclei reached that equilibrium first and the most complex ones did so later, and therefore the time of appearance of species in the fossil record is inversely related to complexity. Species persist as long as the genomic material is maintained in that metastable state and slight shifting across the bottom of this conditional energy valley may cause members of a species to grow larger, smaller, to develop or lose appendages. But essentially they will stay what they are for their whole existence on earth, which varies from about 5 to 500 million years for animals to 3.5 billion years for microorganisms. One may speculate that this metastable state is essential for the persistence of a species and that, when after millions of years the genomic material drops into an absolute energy valley, extinction will ensue regardless of catastrophes!

Thus, the genome goes from the incipient state to childhood, adult life, senescence, and death, and as it does it expresses itself in different forms. Childhood would be the time when the genetic mechanisms develop, and adulthood would result in an escape from the single cellular state to the production of animals and plants. Genomic senescence (the absolute energy valley) may be an inescapable phenomenon that will eventually lead to the extinction of species.

To be effective the genomic configuration must have a consequence in terms of proteins that provide structural stability or catalytic activity. Beyond the well-known concept of the dependence of protein structures upon linear genomic coding sequences there may well be a relationship that makes proteins produced within one cell compatible with each other. This difficult concept requires some reflection upon the function of structural features of macromolecular surfaces. The argument is based upon the suggestion that interacting coding sequences might give rise to interacting proteins (the most simple example). This prediction is an educated speculation and the evidence that can be cited in support is very scarce. Attempts have been made to find such a relationship between receptor/ligand pairs.[12, 13] The new methodology, called proteomics, may promote discovery of relations that may be relevant for this proposal.[14] It is not impossible that the initial DNA structure induced organization and thus loaded the dice for the nuclear refinement process a little different in each case.

So much for a period that is generally ignored as the most uneventful phase of evolution. Of course, most of what has been said about the interregnum is speculation, induced by the Genomic Potential Hypothesis of evolution. True, the skeleton of facts concerning this period has been fleshed out which is the normal function of a hypothesis. Evidence for the existence of bacterial life 3.5 billion years ago is, however, unequivocal[3] as is the fact that larger cells were found 2.2 billion years ago[15] (the time appears to be pushed back with every new report). The appearance of macro-organisms in unimaginable numbers and variety during the early Cambrian period is also well documented in the fossil record and that leaves one no choice but to view the development of macro-organisms as the display of the configuration that the genomes had perfected during the eon of single cellularity; it was the egg that evolved.

There it is, the story of the main events in evolution, shrouded in nebula as is the black hole at the center of our galaxy. We can know it is there by the way it pulls in stars and gases as it bends space into a whirlpool of gravity, and we know that the evolutionary events were there by the way they spew finished creatures from an invisible past without much warning and with their origin hidden from our view.

## References

1. Senapathy P. Independent Birth of Organisms. Madison: Genome Press, 1994.
2. Lopuchin AS. Structures of biogenic origin from early Precambrian rocks of Euro-Asia. Orig Life 1975; 6:45.
3. Schopf JW. Microfossils of the early Archean apex chert: New evidence of the antiquity of life. Science 1993;260:640.
4. Pennisi E. Static evolution. Sci News 1994; 145:168.
5. Stanley SM. Earth and Life Through Time. New York: Viking Press, 1986.
6. Brocks JJ, Logan Ga, Buick R et al. Archean molecular fossils and the early rise of eukaryotes. Science 1999:1033.
7. Schwabe C, Warr GW. A polphyletic view of evolution: The genetic potential hypothesis. Persp Biolog Med 1984; 27:465.
8. Celander DW, Cech TR. Visualizing the higher order folding of a catalytic RNA molecule. Science 1991; 251:401.
9. Cisne JL. Trilobites and the origin of arthropods. Science 1974; 186:13.
10. Gould SJ. Wonderful Life. New York: W. W. Norton, 1989.
11. LaChapelle E. Field Guide to Snow Crystals. Seattle: University of Washington Press, 1969.
12. Goldstein A, Brutlag DL. Is there a relationship between DNA sequences encoding peptide ligands and their receptors? Proc Natl Acad Sci USA 1989; 86:42.
13. Fassina G, Zamai M, Brigham-Burke M et al. Recognition properties of antisense peptide to Arg-8-vasopressin/bovine neurophysin II biosynthetic precursor sequences. Biochemistry 1989; 28:8811.
14. Borman S. Proteomics: Taking over where genomics leaves off. C&EN 2000.
15. Knoll AH. A new molecular window on early life. Science 1999; 285:1025.

# CHAPTER 9

# Thoughts on Multicellularity: How Nature Got Around Darwin

Reproductive success is the ultimate criterion for survival in the Darwinian paradigm, and since micro-organisms are undisputed champions of reproduction the countless creatures that condensed in the shallow Cambrian waters simply drained the old model of credibility. Every living shape appearing at once, crawling, swimming or slithering away from the line that separated the microbial world from the age of large, short-lived creatures, was testimony to an unstoppable drive to complexity with merely adequate reproductive levels. How did nature get around Darwin? The answer may seem extensive, but I simply do not know enough to make it short.

The sharp line between the animate and the inanimate worlds and the speed of transition reflects the powerful force of atomic structure. It took about six times as long (3 billion years) to produce macro-organisms from single cellular life than it did to produce life. Most students of evolution find this quite surprising because, at the level of biological processes, it would seem easier to produce a large animal from a cell than to produce a living cell from elements. In the chapter on the origin of structure this observation is explained by the logarithmic increase in the number of connections required with every additional layer of complexity.

Animals appeared in the Ediacaran, slightly ahead of the Cambrian but most of them were soft-bodied and did not survive except the jelly fish. The first 10 million years of the Cambrian, now known as the Tommotian, brought about a bewildering array of micro-fossils, none larger than a few millimeters. A quote from Stanley's text is illuminating in that context.[1]

"Also found in Tommotian rocks, however, is a host of strange skeletal elements that cannot be assigned to any living phylum and that seem to be unrelated to any group of fossils found in rocks younger than Cambrian age.

The development of the types of skeletons that characterize Tommotian faunas constituted a major evolutionary event. Although skeletons are known to support soft tissue and to facilitate locomotion, such adaptive functions cannot explain why so many different kinds of skeletons developed suddenly in the early part of Tommotian time. It has been suggested that a chemical change within the oceans triggered the production of these skeletons, but this hypothesis does not explain why some skeletons were composed of calcium carbonate and others of calcium phosphate, two compounds with quite different chemical properties. The rapid evolution of various kinds of external skeletons is probably in part attributable to the fact that animals needed protection from enemies; the first multicellular animals must have fed on single-celled creatures and might also have fed on larger plants."

The reader will notice the helplessness of the old ideas facing the prodigious chemistry that gave rise to so many forms during the time of biogenesis. Sensitized by the Genomic Potential Hypothesis one will notice the philosophical faux pas in the statement that the need for protection was enough to provide protection! The observations listed by Stanley (above) contribute to the proposal that the first 10 to 15 million years of the Cambrian were neither a connection to the Ediacaran nor did they give us a preview of the animals that burst onto the scene during the mid-Cambrian. The Precambrian was not a delay but rather the period when the wiring was laid out for macro-organisms, the time of genomic development. Cambrian animals appear to have come each from its own line of pro-forms that collectively were unrelated to either the Ediacaren or the intervening Tommotian.

With the exception of micro-organisms, the cells then and now are not the same; they are separated by 3 billion years of genomic refinement. This genomic evolution, the development of the complexity that led to macro-organisms, was driven by the thermodynamics of DNA/RNA chemistry towards an equilibrium point which was stable enough to persist.

The step toward multicellularity means conquering a different dimension for living systems. Wrong phrase! Life was pushed into different configurations and the new space by the production of new molecules as more coding sequences were exposed within the reorganizing nucleus. Thus, "forging ahead to complexity" really occurred in front of a thermodynamic broom and had nothing to do with selection.

Proteins were synthesized during these nuclear reorganizations, among them perhaps a variety of molecules that would lodge in the cell membrane. These proteins are called CAMs (for cell adhesion molecules) such as integrins and cadherins.[2, 3] Thus, from within the early genome the molecules for higher level organization are produced that would at some later time lead to selective aggregation of cell types.

The association model suggests that colonies began communicating with each other via a contiguous environment (a puddle) by exchanging their metabolic products. Take the case, for instance, where colony A produces a by-product that is eagerly metabolized by colony B and which in turn excretes a metabolite that helps colony A to overcome a metabolic bottleneck. We can extend the picture to include 50 or more colonies in a small space that would exchange end-products so that primary ovum formation might have occurred buffet-style under the influence of chemotactic agents and cell-surface recognition. By this time most readers will have converted colonies to organs connected by a stream (blood) from which cells live by give-and-take to support the extravaganza of a central nervous system that directs the whole scenario with molecules added to the stream in miniscule amounts.[4]

Why would cells listen to such signals from fellow cells at the periphery of an aggregate? Researchers have uncovered suppressor genes such as the retinoblastoma (RB) gene which controls cell proliferation during mitosis.[5] If this gene is modified by a mutation the cell will grow out of control (cancer, or better return to the wild type) whereas greater abundance of the gene product causes cell death (apoptosis). One can hypothesize that cell lines could have developed into single-purpose colonies and several of these colonies, coming together, could have gone through many cycles of fusion and separation, and finally come to rest in a unit, forming a federation of talents wherein cell types could ply their trade as filter feeders or gas exchangers. They were quasi-transferred into the milieu interieur of an organism under the rule of a set of master cells. Such box-in-box scenarios have an esthetic quality that has always been attractive but not necessarily true. Consider, however, that organs of an animal, such as liver, kidney, brain, are more like the same tissues in another animal then they are like different tissues in the same body.

It is also conceivable that cell lines could have developed all the chemical intricacies that would, in proper relation to each other, express all of the organs and the structural components of a macro-organism, guided only by genomic rearrangements to produce a stem cell.

Collectively, science knows nothing about the topic of this chapter. The objective of this discussion is to formulate problems that can stimulate research and to illuminate the concepts. Still, the various factors mentioned are real. CAMs have been discovered recently,[2] growth factors have been known for a long time,[6] and, of course, hormones are in principle no different than the pheromones whereby insects communicate over long distances.[7]

Chemotaxis is another reality and chemotactic agents are important for many processes in the body.[8] Thus the endocrine chemosensor system might have built the organisms by pulling cells together in a staging area as opposed to the contemporary belief that organisms have developed an endocrine system.[4]

This goes beyond the proposal attributed to Peter Medawar who suggested that "Endocrine evolution is not an evolution of hormones, but an evolution of the uses to which they have been put".

To develop an endocrine system of the complexity and specificity as observed in contemporary animals by a chance-oriented mechanism is not an option. The probability of simultaneous coordinated mutations that led to the production of a messenger in one part of the body and to the production of the receptors with the appropriate tissue response and distribution in another part, is unqualified nonsense.[4] From this realization evolves the condition that interacting systems have to be developed in proximity at the nucleic acid level. The endocrine system components have, no doubt, evolved during the development of the nucleic acid core in the primordial pool; the associations that developed much later represented merely a process of selection based on chemical recognition. The organisms that stayed single were formed from genomic material that did not contain the messages for one or the other recognition factors such as nuclear organization factors,[9] cell adhesion molecules or hormone receptors.

Genomists would reason that, as a byproduct of such a process of organo-synthesis, one should find cells with a partial complement of these factors, perhaps not enough to allow organization of macroorganisms, but enough to form temporary or low affinity associations. Some single cellular eukaryotes do appear to have steroid receptors,[10] and some prokaryotes have specific binding sites for collagen,[11] but they do not have the full complement of proteins that would allow for the development along multicellular lines. In this context it is interesting that interaction with matrix proteins (collagen, proteoglycans, fibronectin and laminine) is required for organosynthesis, i.e., the formation of lobulated secretory glands, for example, will not occur without the interstitial components. Interaction with ground substance requires specific receptors as organizing factors.[12, 13]

The new hypothesis leads to the prediction that the coding sequences for hormones, for example, are very old, that they were produced in the primordial pool. Their expression in one organism versus another is a mere problem of association of these coding sequences with a developing focal point so that a certain molecule will not necessarily end up in the same species or even in the same phylum. This prediction too appears to be true in that vertebrate hormones are found in invertebrates,[14] animal hormones in plants and single cellular organisms, and even in prokaryotes.[15] I want to transmit the idea that evolution may have been a matter of organization of the products of primordial chemistry into more or less efficient biological units, and that there is no room for adaptation.

Various degrees of association can be observed between multi- and single-cellular existence. Some life forms have it both ways like bacteria that occasionally can form predatory units or very elaborate fruiting bodies, and slime mold goes through a regular cycle of single cellularity and multicellular fruiting stalks. The signal for association is a small molecule that diffuses through the medium and which is different for different subspecies of slime mold. If one mixes the single cellular stage of two different species, and places a drop of cyclic adenosine mono-phosphate (AMP) into the medium, one of the two species will migrate toward the higher concentrations of cyclic AMP and will start forming fruiting bodies.[16] Here then we have a complete replay of the association scenario presented earlier except that in the case of the slime mold the association is reversible and does not lead to the formation of larger, permanent bodies. Noncellular slime mold also produces different stages, one of which has been compared to a large muscle cell.

Bryozoa form colonies of astounding variety and, in some instances, even mobility (*Cristatella mucedo*). Individuals (autozooids) of bryozoa colonies appear to be capable of differentiation into heterozooids of fantastic shapes.[17] One of the appendices of *bugula* looks like a bird's head and beak and is, in fact, capable of turning and snapping at intruding objects.

The protozoan *Myxotricha paradoxa*, a complex symbiont that lives in the gut of certain Australian termites, plays in turn host to three symbiotic forms of life, namely small spirochetes, large spirochetes (for propulsion), as well as bacteria that mediate the digestion of wood fiber in the digestive area of the protozoan.

Among social insects there exists an even looser association between different units. An anthill should actually be considered a single organism and the individual ants are organs that are ruled and directed by preprogrammed responses to pheromones. The size of these organisms varies from a few grams to several kilograms. A single colony of the African driver ant *Dorylus wilverthi* may contain about 20 million workers and weigh up to 20 kilograms.[18] The individual members of an ant colony are haploid, meaning that the individual ant contains only one full set of chromosomes as opposed to the two sets contained in normal cell lines in macro-organisms. Thus it is ever more tempting to look upon such an aggregate as an individual without a skin. The soldier cast would be equivalent to the macrophages and perhaps the immune cells in our body, and they are equally preprogrammed to do what they are doing regardless of the outcome of their action.

How did we get here and why? The master cell that slips a control unit into all cells to make them subordinates appears again in the form of the queen of the nest. The genomic program that makes ants do what is expected upon contact with pheromones comes from the only diploid in the colony, the master cell, the queen in this case. It is one of the connections that proves nothing *per se*; it certainly entertains and stimulates one's imagination but it also sets the limits for the whole argument of multi-cellularity.

All of these relations that are evident in living systems have to have been contiguous at some time. Contiguity is a conditio sine qua non for any theory of science. The Darwinian 'point origin'[19] is a concession to contiguity albeit, at the price of adopting a miracle (single origin). For the GPH the generic basis of life is nucleic acid produced with endless permutations over nearly the whole surface of the earth. The origin of all forms of life is contiguous through the properties of nucleic acids, though not necessarily by physical proximity. The contiguity of the origin of life is therefore not limited to the earth but is universal in the true sense of the word and for that reason only a genomist (a name for those to who accept the premises of the GPH) would be entitled to predict that life should exist on other planets and that it would be in principle like ours. Since evolutionary advances can only come from the nuclear material, they would come from nuclear material everywhere and therefore the next layer of complexity would be similar again; differences will appear toward the periphery (see Chapter 11). This nuclear material is the substratum for cells and by extension for organisms.

In as much as organisms are a permanently reproducing unit, at least one cell line, the stem cell, must have all the information that causes colonization. In the association model a temporary amalgamation of nuclei, a syncytium may be an intermediate step which divides again into tissue cells after branding the cells as members of an organization somewhat like the thymus gland identifies endogenous (as opposed to foreign) proteins in juvenile forms. As the master cell divides, control of the somatic cells gets to be less stringent, and ultimately the cells sort themselves into groups according to their cell surface association proteins (CAMs) and resume their precolony activity, now however under stringent growth control as kidneys, liver, muscle and so on. Their activities are steered and their growth is limited by factors that emanate from the master cells and circulate together with a continuous stream of nutrients and act upon, for example, the retinoblastoma gene.

Once macro-organisms had developed they were subject to functional selection which amounts to a 'thumb up or down' decision without a second chance. It appears, however, that the selection rules allowed the formation of the first organisms against the disadvantages of being large. With these conclusions comes a set of problems to solve for future generations, problems pertaining to the transition from the chemistry that builds structures to one which repairs and controls structures. I am not referring to experiments inspired by the neo-Darwinian model which purports that new activities are derived from gene duplication products. The test for the repair of defects was actually performed by artificially introducing an error into an enzyme of a micro-organism and to watch for spontaneous repair by 'targeted' mutations[20-22] which ended, as it had to, in failure. Chemistry, amazing as it is, does not see a big picture, the benefit of the formation or undoing of bonds. Yet, the whole neo-Darwinian model is build upon the assumption that after gene duplication the new gene will mutate its way into a new and needed function.

The postulates of the Genomic Potential Hypothesis proffered in this chapter are in line with the most common way of existence for animals today which depend upon plants, plants and animals need bacteria, bacteria need hosts, and so on. Life at large is a single phenomenon, and life forms are merely a fluctuation of this phenomenon to yield more of one form at certain times and other forms later.

Evolution must be genomic exclusively because there is no pathway, within the extent of our understanding of molecular biology, that could feed back into our genome to effect changes that would make us more resistant to natural phenomena. Our brain, which makes us more resistant to the environment, was developed by genes but not by environmental vacillations.

Multicellularity, single cellularity, the difference is genomic complexity, nothing changes fundamentally as life proceeds from one to the other.

## References

1. Stanley SM. Earth and Life Through Time. New York: Viking Press,1986.
2. Albelda SM, Buck CA. Integrins and other cell adhesion molecules. FASEB 1990; 4:2868.
3. Chen Q, Kinch MS, Lin TH et al. Integrin-mediated cell adhesion activates mitogen-activated protein kinases. J of Biol Chem 1994; 269:26602.
4. Schwabe C. New thoughts on the evolution of hormone-receptor systems. Comp Biochem Physiol 1990; 97A:101.
5. Cobrinik D, Dowdy SF, Hinds PW et al. The retinoblastoma protein and the regulation of cell cycling. TIBS 1992; 17:312.
6. Burgess WH, Mehlman T, Friesel R et al. Multiple forms of endothelial cell growth factor. J Biol Chem 1985:11389.
7. Henzel WJ, Rodriguez H, Singer AG et al. The primary structure of aphrodisin. J Biol Chem 1988; 263:16682.
8. Bonner JT. Chemical signals of social amoebae. Sci Am 1983; 248:114.
9. Cohen M, Lee KK, Wilson KL et al. Transcriptional repression, apoptosis, human disease and the functional evolution of the nuclear lamina. TIBS 2001; 26:41.
10. Loose DS, Schurman DJ, Feldman D. A corticosteroid binding protein and endogenous ligand in *C. albicans* indicating a possible steroid-receptor system. Nature 1981; 293:477.
11. Switalski LM, Ryden C, Rubin K et al. Binding of fibronectin to staphylococcus strains. Inf Immun 1983:42.
12. Aumailley M, Nurcombe V, Edgar D et al. The cellular interactions of laminin fragments. Cell adhesion correlates with two fragment-specific high affinity binding sites. J Biol Chem 1987; 262:11532.
13. Hoffman MP, Nomizu M, Roque E et al. Laminin-1 and laminin-2 G-domain synthetic peptides bind syndecan-1 and are involved in acinar formation of a human submandibular gland cell line. J Biol Chem 1998; 273:28633.
14. Georges D, Schwabe C. Porcine relaxin, a 500 million-year-old hormone? The tunicate *Ciona intestinalis* has porcine relaxin. FASEB 1999; 13:1269.
15. Schwabe C, LeRoith D, Thompson RP et al. Relaxin extracted from protozoa (*Tetrahymena pyriformis*). J Biol Chem 1983; 258:2778.
16. Bonner JT. Hormones in social amoebae and mammals. Sci Am 1969; 220:78.
17. Margulis L, Schwartz KV. Five Kingdoms. New York: W. H. Freeman and Company, 1988.
18. Wilson EO. Sociobiology. Cambridge: Belknap Press of Harvard University Press, 1980.
19. Mayr E. Darwin's influence on modern thought. Sci Am 2000:79.
20. Burbaum JJ, Raines RT, Albery WJ et al. Evolutionary optimization of the catalytic effectiveness of an enzyme. Biochemistry 1989; 28:9293.
21. Blacklow SC, Knowles JR. How can a catalytic lesion be offset? The emergence of two pseudorevertant triosephosphate isomerases. Biochemistry 1990; 29:4099.
22. Blacklow SC, Liu KD, Knowles JR. Stepwise improvements in catalytic effectiveness: independence and interdependence in combination of point mutations of a sluggish triosephosphate isomerase. Biochemistry 1991; 30:8470.

# CHAPTER 10

# Natura Non Facit Saltum:

## Nature Does Nothing in Jumps

Latin is intimidating, and people tend to defer to anything expressed in that venerable language. Of all the errors chiseled into our cultural foundation in Latin, the title of this chapter is remarkably wrong. The creation of atoms, the movement of electrons, molecular energy levels, everything is quantized and h (Planck's constant) represents the energy difference between nearest energy states anywhere in the visible universe. We live by feeding the energy difference between ground state and excited electrons into our system, taming the electron on its way down, so to speak, draining it of its energy one small package at a time. The sub-atomic world is rough but, unlike bacteria and smaller creatures that are buffeted by Brownian motion, metazoans do not feel the roughness. Macroscopic life feels molecular action as temperature, which seems to be a step-less quantity. The world perceived through our senses seems smooth and continuous so that we have to restore some credit to the idiom that is the title of this chapter.

Evolution, is it smooth or sporadic and does it matter? Sometimes our big problems boil down to silly questions. The evidence is right before us, could one not just recognize saltation without much ado? The problem is more complicated because the evidence comes in the form of still pictures (fossils) that need to be assembled to make a story and, because frames are taken millions of years apart, there is latitude for different hypotheses to exist. The mode of evolution, i.e., smooth and imperceptible, versus stop and go, versus rapid appearance with minor adjustments, are questions that are deeply embedded in different hypotheses. The answer to these questions determines what hypothesis of evolution will be viable in the final analysis so, does nature make jumps in the macroscopic world?

Darwin thought that organisms would change imperceptibly as a function of time, but that idea was based upon a mechanistic model that had been inspired by the ways of animal husbandry. The fossils, however, have always shown an abruptness at the incipience of species that was usually explained by an incomplete fossil record. The intermediates that were needed to smooth out the relation between reality and model have remained elusive. Under the pressure of data, a group of Darwinians broke rank with the mainstream and designed a scheme that would combine periods of stasis with spurts of rapid changes called punctuated equilibrium.[1] The gentle evolutionary slope becomes a staircase that would better match the sudden appearance of new species in the fossiliferous layers. The trouble is that modern science does not provide a mechanism for such rapid changes in a well-established species. Mayr explains that isolated groups, that have escaped notice in the fossil beds, suddenly produce offspring with advantageous features that brought reproductive success and visibility.[2] This explanation (founder hypothesis), like the punctuated equilibrium variant of Darwinism, is based upon an unknown mechanism ("motley egg" mechanism). These authors are searching for a way to incorporate into the standard model observations that are the core of the Genomic Potential Hypothesis wherein this apparent abruptness is a natural consequence of development.

According to the Genomic Potential Hypothesis, cells produced at the lower Archean slowly and continuously rearranged their nuclear material, streamlining it to satisfy energy minimization. At the beginning of the Cambrian and during later periods, efficiency caused the rate of growth to increase and cells began to display binding proteins so that they did not separate before the next duplication. That caused their metazoan gene configuration to be activated (metamorphosed) such as to produce phenotypes. Each group of pro-forms would give rise to one group of animals, for example, and pro-forms of increasing complexity would ripen in every major period giving rise to more complex creatures up to the Quaternary when the supply of 'stem cells' was exhausted. In this scenario nothing ever changed into anything else, and that concept reasserts itself in the invariability of species including their proteins.

Shifting attention back to the developing cells we can think of complexity as a latent form of energy (entropy). For each degree of potential complexity (amount and variability of nuclear material) in a primary cell, a corresponding amount of organizational complexity is required before the primary ovum is ready to jump into macroscopic existence. Unlike the electronic transition where all suboptimal energy is wasted, organizational complexities (control genes) accumulate and grow slowly until they satisfy the developmental potential of the genomic Anlage of a primary stem cell. Thus, if the stem cell has developed the genes to produce lungs and legs, it must also develop segmentation to distribute these functions sensibly. This image gives form to the symmetry and complexity numbers in Chapter 8. Homo came very late in this game because there was a lot to be organized. Genomists conclude that the genome develops slowly and continuously and that the phenomenological phase of evolution is sudden but continuous as well, sort of like the transition of liquid water to ice.

The succession of many transition events has created the illusion of succession among species for Darwinians but denied them the intermediates that are required to prove the point. These gaps in turn provide the most powerful evidence for the concept of polyphyletic evolution. The equation of state for the Genomic Potential Hypothesis-type transition is literally pressed into the layers of the Cambrian stones for everybody to see. The time frame of deposit rules out interspecies conversion but rather points to many phase transitions, one for each débutante at the threshold of metazoan life.

If this seems at first to be a difficult concept at least someone else apparently saw it and I can always shed monastic inclinations for a moment of comic relief. Bertrand Russell's book *Why I am not a Christian* inspired me to write a short paper with the title "Why I Am Not a Darwinian" for *The Scientist*. Here is a ditty produced by a referee at the sight of my remarks " I don't see anything unconventional or novel in the submitted assay. It is a standard presentation in a standard form of a perspective that is no secret". Many of my readers may remember the three stages of discovery attributed to Pasteur according to which nobody believes it at first, then everybody knew it all along (see above) and finally reassignment of credit.

The presentation of phenotypes was a sustained fireworks of astronomical dimensions that capped a stretch of developmental time equal to about 1/5th of the age of our universe. It was an eon of genomic evolution guided by free energy and entropy rather than by that deadly Darwinian clinch for survival. A small fraction of the cells, that propagated for 3 billion years and passed their genomic refinement along perhaps 3 times a day or 1000 times a year for a total of $10^{12}$ generations, went off in clusters and dimmed again at various times between the Cambrian and the Quaternary. New ones lit up in every major period until at last homo arrived to start this investigation and to finish it before his light will dim as well.

The fossil record that is so troublesome for the Darwinian model actually would support saltation scenarios. In fact, Mayr's lone founder idea and the punctuated equilibrium model rely upon what I call the 'motley egg' mechanism (motley in this context means that the egg has genomic spots that differ from the parent gene). Neither the founder idea nor the punctuated equilibrium model would work without an egg that would in some way vary beyond the Mendelian rules. Rewriting the history of the evolution of amphibians and reptiles one might start with *Eryops*, the oldest known amphibian from the Pennsylvanian and ask whether it laid an egg from which the first reptile *Hylonomus* emerged and then another one to produce *Diadectus*, a large herbivorous reptile? During the next season then, because all came on board at about the same time, might those eggs have given rise to members of the Pelycosaur group such as *Dimetrodon* from the lower Permian? Of course the description could be a bit more complicated, but the fossil record in essence invites this conclusion and it is impossible to categorically deny it. Rejection must be based upon our lack of experience with transmutating secondary eggs (eggs from animals), upon our knowledge of genetic stability and control mechanisms and finally upon our knowledge of the complexity of integrated pathways in metazoan. That is a serious drawback for any model even in biology where the reins of epistemology seem to hang loose. The desire for simplicity must be curbed as Einstein suggested with subtle Germanic sarcasm: "Science should be as simple as possible, but not simpler".

Modified egg ideas such as punctuation and the lone parent idea of Mayr all state identical notions although the proponents of animal to animal evolution make no point of it. At one time an ape, for example, would produce an intrauterine egg which, if fertilized by an ape spermatozoa, produces an offspring with prehuman potential. However little change one wants to award the first member of a new taxon, it must be fixed in the gene and significant enough to open the road to humanity. This proposal is essentially a variant of the reptile-producing amphibian egg idea, less facetiously presented. The fossil record rules out the small-step conversion but not the egg mutation hypothesis which does not require intermediates but runs into problems with biology.

Summing up, it seems that the saltatory appearance of the fossilized species in the past does not support the Darwinian picture. It is consistent with the "motley egg hypothesis" and with the Genomic Potential Hypothesis. The difference is distinct insofar as in the Genomic Potential Hypothesis a tiger has never been anything but a tiger, once in the genomic form within a cell and, thereafter, in a fur.

All hypotheses of evolution force us to accept transitional forms of some kind of living, feeding, and reproducing eggs that would survive 3 billion years to metamorphose into a living reproducing animal. Smoothness of phenotype transitions is not observed and all other explanation comes at a price.

It is not that everybody has given up on finding intermediates. Speculations have been numerous, but whenever evidence became subsequently available the correlation began to fail. A highly publicized example involves coelacanth (a lobe-finned fish) and *Ichthyostega* (an amphibian). A prominent Darwinist writes as follows, "The lobe fin itself is formed of an array of bones resembling that found in amphibians; similarly the complicated teeth of lobe-finned fishes closely resemble the teeth of early amphibians. These features alone strongly suggest that amphibians were derived from lobe-finned fishes, but additional features make the derivation a certainty. *Ichthyostega* had four legs as do amphibians, but its skull structure was remarkably like that of a lobe-finned fish. The creature had also a fish-like tail—one feature of its ancestors that was probably of no use on land. Because of this intriguing combination of features, *Ichtyostaga*, which was not discovered until the present century, represents what is commonly termed a 'missing link'. [3]

Here their luck ran out because the living Coelacanth was rediscovered in the deep waters around the Comoro Islands in the Indian Ocean[4] and not in shallow waters where they could practice walking skills. The fish that had been credited with ancestorship of the vertebrate amphibian *Ichthyostega* had merely a soft notochord instead of a vertebra. The development of a vertebra is not a minor matter when, in fact, this bone had become the basis of a major phylum. Coelacanth turned out to be ovoviviparous. This makes coelacanth the more advanced creature and leaves *Ichthyostega* without a taxonomic parent. It leaves us with two more examples of nature's disrespect for our prejudices and in a practical way with a sense of the futility of homology arguments across species. Without a living Coelacanth the notion of an ancestral relationship between these two species would have been hard to dispel because it seems that the modern evolutionists have forgotten the difference between dependent and independent variables.

At the end of this analysis of natural jumpiness one has to concede that nature is continuous at the macroscopic level. It boils down to the fact that the jumpiness of nature is a matter of the mode of observations and the title is wrong only in that it excludes what was not known when someone mused about the visible world.

## References


1. Eldredge N, Gould SJ. Punctuated Equilibria in Models in Paleobiology. New York: Freeman, 1973.
2. Mayr E. Biological classification: Toward a synthesis of opposing methodologies. Science 1981; 214:510.
3. Stanley SM. Earth and Life Through Time. New York: Viking Press, 1986.
4. Fricke H. The fish that time forgot. Nat Geo 1988:825.

# Chapter 11

# The Invariance Concept

Tucked away, if it were possible, within the hollow of the tip of a dart gun needle aimed at man or beast one could make a memorable discovery. Just a moment after skin penetration the overwhelming impression would be 'sameness'. A human, a pachyderm, a mouse, a rat, donkey, ape, in all cases the first experience is the epidermis, then the dermis, a layer of fat and collagen followed by muscles and sometimes a hard landing on a bone. If the needle were fine enough to penetrate the cell membrane and the nuclear envelope, the likeness would stretch to almost all of life because one would see strings of ATCGTACGCGG—as far as the eye reaches. How can the core of everything be so alike when creatures are as different as elephants and humming birds or tunicates and pigs? By reflex most biologists will point to a common ancestor that lived during the Hadean.[1,2] The new model suggests that the common ancestor did not live but that the universality of the physics that underlies chemistry had played the role of the pluripotent ancestor.[3] The number of possible chemical compounds is nearly limitless which raises questions as to how fundamental mechanisms could have come out so uniform.

It is a peculiar type of restriction that dominated the biogenesis scene. The first phase is carbon chemistry of the Miller-Urey type which led to the production of nucleic acids and amino acids among other molecules.[4] Nucleic acids were the gate, the nozzle through which everything that would be part of life needed to pass and from experience we know about the tendency of chemistry to favor, above other compounds, the production of these biomolecules. Passive selection of constituents continued, guided only by what is thermodynamically favorable, until the first stages of self-replication became important. It does not take much imagination to see that nothing had a chance in competition with nucleic acids. Other compounds formed crystalline structures of billions of molecules but without the potential to produce informative defects. Errors in crystalline structures are too rare to carry the amount of information that living systems require.[5] In contrast, the structure of nucleic acids is semi-crystalline but has an error (information) at every level of the helix. Life is built from the "memory out" toward the periphery and since the memory always reads [(A)x (T)x (C)y (G)y ] the nucleic acids in widely separated origins did not look much different. Nucleic acids work like the monotonous binary code tape of a "Turing machine", a reading device that gives patterns of zeros and ones to which a meaning can be attached only after one decides what the blocks of symbols are to represent. The four bits of our nucleic acids, A, T, C, G provide a greater information density than a Turing tape and therefore a much shorter DNA string is sufficient to spell out a structure so complex that the most refined feature of the living world, the human brain, cannot comprehend it.

Polymerization of nucleotides was the founding event, the " big bang of biology" which led to the self-perpetuating structures. Nucleotides were the last representatives of the "forgetting" world; polynucleotides are memory bits of significantly greater information density than silicon chips, they represent the ultimate nanotechnology. Once polymerization had begun the chemical pools became ancestors. The polymers will have been different in detail but in every location themes of the primary polymers plus variations would have been recognizable if one were to analyze the nucleotide sequence of each pool. One could mentally follow the world line of each of the biogenic pools to the time of fruition and note that each would evolve different species and variants of multicellular life and others, perhaps the majority, would remain single cellular.

Presently the "dart needle perspective" makes sense. The basic material must have been the same to make a living creature. The early conditions that led to life are as restricting as the apertures the physicists use to analyze radiation except, as biology progresses, the holes get wider. The dart needle and its passenger, penetrating the skin, the cell and finally the nucleus, traveled backwards in evolutionary time through the apertures to the absolutely required initial structures, the nucleic acid core.

Proteins, the catalytic surface of life as well as the structural framework, represent a derived complexity, which is accessible only via the nucleic acid level of organization. Thus, when thinking of the multiple origin concept one must think of the nucleotide level information as a keyboard of genomic material which, if played by proper control mechanisms, can give rise to different organisms. The control mechanisms in turn are the products of intragenomic chemistry and the first layer of subordinate control may be the homeo genes that control protein expression along the spinal axis in vertebrate animals and in segments in invertebrates.[6-8]

Proteins are unstable yet their production is an inevitable step on the way from nucleic acid to life. While there is latitude in terms of the kind of proteins acquired, as concerns the functions of these proteins there are significant restrictions. Their catalytic activity must be supportive and not destructive for the parent organism. All conditions are de facto rectifiers that tend to spread uniformity.

The triplet code gives life to the bio-memory structure but again without any overt wisdom. One must leave open the possibility of a deeper and not immediately obvious connection between nucleic acids and L-amino acids[9] that might have helped to establish reading breaks and on-off signals.[10] These are deep waters to cross but, once the hypothesis has gained some ground, investigators will find it worthwhile to examine the nuclear reactions for a structural rationale.

Whatever the outcome one can be sure that a physical contact type selection has led to the genomic code. Not only does the new hypothesis predict such contact, the status of evolution as a discipline of science depends upon it. Frozen accidents are not the stuff of hypotheses but more likely failures of insight. This means that the principle of code development is discoverable and with it, slowly to be sure, the whole mechanism of biochemical evolution.

Clonal development of life means condensation of pieces of memory from a pool of limited volume until enough functions can be read from the nucleic acids core to support autonomy. A smooth transition of complex chemistry to complex biochemistry must proceed through stable (immortal) steps and products that last until the next

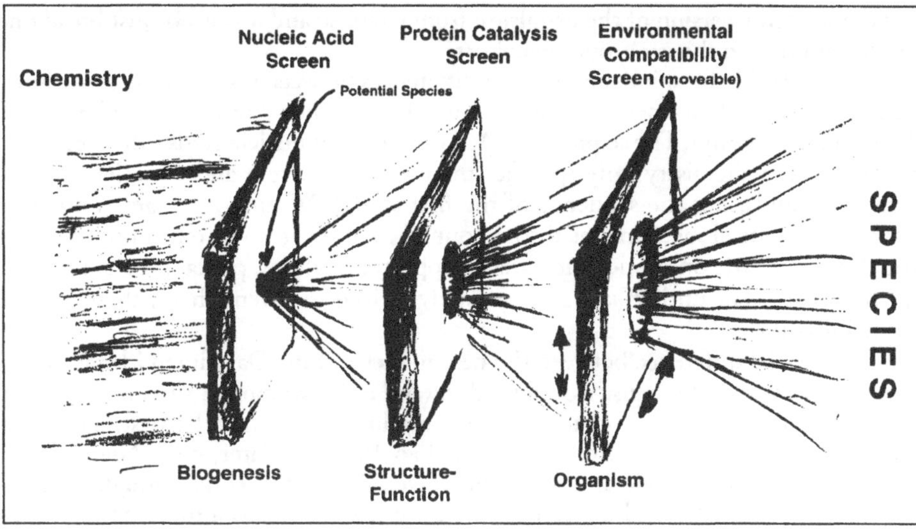


Fig. 11.1. This figure illustrates the process of selection not by restriction but by fit into the next hierarchy of structures. The suitability is a purely chemical phenomenon and the pinhole plate is an imperfect symbol of fit of fragments into the structures of nucleotides which, in turn, are swept up by polymerization reactions. The aperture at the next station restricts on the basis of suitability of nucleic acids for the development of catalytic units. The aperture symbolizing that differentiating process is larger because there are many ways for nucleic acids to end up in living systems in contrast to the first aperture in barrier 1 that is restrictive to the point that only nucleic acids can cross that threshold. The third aperture is the largest yet because there are many ways to make a living. It is movable because the living units that appear will be sensitive to environmental conditions that can mean failure to life forms which would be quite suitable to exist under different conditions.

addition spark the processes of life. The prediction that these 'assembly functions' should show a great deal of similarity among the majority of life forms is based upon the simple argument that in biogenic pools the constituents must have been similar within limits and that significant differences are only recognized when the units start to metabolize. In spite of different potentials (DNA sequences) the first expression of life was probably rudimentary which is why the survival functions like energy storage, locomotion, and support look today as we see them from our perch in the dart-gun needle. All additional complexity is organizational rather than fundamental, and those organizational levels are presently under scrutiny.[11, 12]

The membrane has the premier function in the process of biogenesis. It allows for individual ownership and retention of biocatalysts, and thereby for up to a million fold increases in catalytic activity. Substrate/enzyme ratios in cells may approach unity and thus enzymes can actually change the equilibrium of some reactions. Clearly, membranes are essential and the hurdle for nascent life is the need for a selectively permeable membrane... that means a membrane that contains, suspended in its lipid layers, the first communication proteins.[13, 14] The cell must breathe at once if there is to be any future and that again equalizes units from different clones. Is it surprising then that all life forms have membranes? Shapeless wafting life is a thing of poor science fiction. Membrane formation is the moment when life became competitive, it was the scientific version of the expulsion from paradise and it was not just bread one was to eat but also the evolutionary cohorts.

Post-assembly functions show larger differences between taxa because those functions are due to clone-specific genomic organization (not due to mutated enzymes that work differently). The sequence of A,T,C and G's in nucleic acids differs between clones because chemistry only specifies the general structure and complementarity between strands, not the sequence of the first strand. Writing messages is done by essentially identical mechanisms throughout life, but what is written is not the same and thus, very early on species and variants appear; kingdoms, phyla, classes, orders et cetera, had to wait until *Homo sapiens* finally arrived to invent them, these are not natural divisions.

The crucial difference between the new model and the Darwinian idea is how it began.[15] Darwin, as biologist would tend to (and did) extrapolate animal husbandry to speciation on one side and to the single origin looking into the past. This view found all but absolute confirmation more than 100 years later when Nierenberg[16] discovered the genetic code and when the universality of this code throughout all of life was noticed. Haley and his comet could not have provided a more powerful endorsement of Newton's theory than Nierenberg gave to Darwin, yet, in the final analysis, what seemed obvious, was also incorrect.

The genetic code invariance persuaded evolutionists to throw the first shy proponents of polyphyletic models into the dungeons.[17, 18] The contention that chance would not have provided even for two origins with an identical triplet codon for protein synthesis was then and still is absolutely correct and certainly relevant, but the conclusion reached because of it by nearly everyone was not. Multiple origins were declared impossible whereas chance should have been disqualified as an inappropriate term in this equation (which it is), and this was the error that has dominated thinking for more than 150 years and is still defended with vehemence.

With many liter-size "bio-reactors" the planet would still have looked arid to alien visitors. The Genomic Potential Hypothesis will have all of them make proteins in essentially the same way because it is the only way for making proteins efficiently and repeatedly and to remember forever how it was done. If other routes were possible we would see them today, just like we see differences between prokaryotic and eukaryotic genome organization. Versatile as it is, chemistry is not resourceful enough to provide two independent pathways to life.

Does it still seem difficult to see a large number of biogenic pools giving rise to species that could eat each other and, within mixing zones, even breed with each other? I am afraid so, our intuition was raised on randomness in chemistry. The new hypothesis is a bit feisty on this argument, suggesting that the products of a new biogenic period, if started under the same conditions and with the same amount of starting material on an earth-type planet, would have landed squarely on the chaotic attractor of our life, meaning the production of all creatures we see now, though the names may be different. This prediction, I am confident, will be confirmed should we be able to find life on a planet remotely similar to ours simply because chemistry is universal. The chaotic attractor of life is shaped by the structure of baryonic matter which is, by all accounts, the same out to the edge of the universe. Yet, at this time one could throw together all the baryonic matter in reach and wait for an awfully long time without seeing any life emerging from the mixture, not because it does not work, but because reaction conditions are a major ingredient in the chemistry of the origins of life.

Can one be so sure about this invariance concept only because we can never test this prediction? Not really. The Jovian moons are within our reach and Mars is still in contention as a life-supporting rock, so one must consider the idea in principle testable. Life will be the same and everybody who is searching for life elsewhere knows it intuitively, because they are looking with the methods it takes to detect our kind of life. They are all unwitting supporters of the invariance concept of the Genomic Potential Hypothesis.

A clonal model places all the discriminating action at the very bottom of the evolutionary events and that proposal, in turn, leads to the uniformity argument. Major development occurs early on when the chemical system is at maximum plasticity, and that is not only the source of basic uniformity but also the major source of emotional resistance to the new model within the scientific community. The proposal that chemistry worldwide would either give rise to the same mode of life or to no life at all seems an assault on common sense, the same common sense that has been so successfully assaulted by all the pioneers in science early in the past century.

Invariance theory is the name Einstein had perceived for what became known as Relativity. As space condenses, everything slows down for an outside observer. Amazingly and totally against one's intuition time passage, as defined by the speed of light, remains unchanged and life will speed up or slow down depending upon the density of space, and we would never notice. No doubt it is true, but it is a "conceptual truth" that cannot be tested on living creatures because gravity in slow-time regions will crush water-containing structures. Experiments with particle accelerators, however, have shown clear evidence of time dilation. Such a conceptual unity envelope also extends across the biogenesis scene. It is this cosmic concept, frozen in atomic structure, that becomes obvious first in the form of nucleic acids, which lets us be certain that life anywhere will rest on the same chaotic attractor as life on earth.[19] While riding the dart needle, we witnessed that unity, the differences that are visible only at greater distance, are organizational primarily and quantitative secondarily; they depend upon the linear arrangement of the nucleic acids.

Science will progress and it is a somewhat oblique tribute to the religious undertow in the evolution of science today that the major hypothesis of our own development has not been permitted to join the high-flying intellectual world of our colleagues in physics, astronomy, and cosmology, even though the concepts in biology are as high-flying, sophisticated and, yes stunning, as those of physics and astronomy. Consider, please, the immense number of genetic code words assembled in part randomly (primary strand) and in part by complementarity. At the sight of such a distribution of bits of information Frank Plumpton Ramsey would calculate that nearly every possible structure should be discoverable in this matrix.[20] The complexity of a design discoverable in large distributions of points (stars or nucleotides) depends upon the total number of points in a set. With a thousand moles ($\sim 6 \times 10^{26}$ nucleotides) conservatively estimated to have been on earth during biogenesis, Ramsey would grant us every conceivable configuration of life.

## References


1. Benton MJ. Diversification and extinction in the history of life. Science 1995; 268:52.
2. Doolittle RF. Similar amino acid sequences: Chance or common ancestry? Science 1981:149.
3. Schwabe C. Evolution and chaos. Computers Math Applic 1990; 20:287.
4. Miller SL. The atmosphere of the primitive earth and the prebiotic synthesis of amino acids. Origins of Life 1974; 5:139.
5. Laszlo P. Chemical reactions on clays. Science 1987; 235:1473.
6. Crawford M. Homeobox linked to gene control. Science 1988; 242:1008.
7. De Robertis EM, Oliver G, Wright CVE. Homeobox genes and the vertebrate body plan. Scientific American 1990:46.
8. Gehring WJ. Homeo boxes in the study of development. Science 1987; 236:1245.
9. Yarus M. A specific amino acid binding site composed of RNA. Science 1988; 240:1751.
10. Bost KL, Smith EM, Blalock JE. Similarity between the corticotropin (ACTH) receptor and a pepide encoded by an RNA that is complementary to ACTH mRNA. Proc Natl Acad Sci USA 1985; 82:1372.
11. Lonai P, Orr-Urtreger A. Homeogenes in mammalian development and the evolution of the cranium and central nervous system. FASEB 1990; 4:1436.
12. Marx J. Homeobox genes go evolutionary. Science 1992; 255:399.
13. Edelman GM. Morphoregulatroy molecules. Biochemistry 1988; 27:3533.
14. Cheng H-C, Kemp BE, Pearson RB et al. A potent synthetic peptide inhibitor of the cAMP-dependent protein kinase. J Biol Chem 1986; 261:989.
15. Schwabe C, Warr GW. A polphyletic view of evolution: The genetic potential hypothesis. Perspectives in Biology and Medicine 1984; 27:465.
16. Nirenberg M, Leder P. RNA codewords and protein synthesis. Science 1964; 145:1399.
17. Nursall RR. On the origins of the major groups of animals. Evolution 1962; 16:118.
18. Kerkut GA. Implications of Evolution. London and New York: Pergamon Press, 1960.
19. Jensen RV. Classical chaos. American Scientist 1987; 75:168.
20. Graham RL, Spencer JH. Ramsey theory. Sci Am 1990; 263:112.

# Chapter 12

# On the Evolution of Humans

Ancient bones are somewhat like wind vanes that show from which direction a particular hypothesis breezes across the fossil field. If a paradigm is useful one should be able to predict what will be found at the end of the projected course. This means that one can put a general evolutionary hypothesis like a grid over the pattern of evidence and see how fossils and expectations match. In the standard model predictions are not possible because the phenotype lattice radiates from one spot (the common ancestor) with every beam studded with chance-initiated branch points that give rise to unpredictable patterns.[1] By rules of the Genomic Potential Hypothesis the future position of a species in the hierarchy of taxa is, in principle, predictable. The discussion of our own past will reflect this fundamental change in philosophy.

The human brain is monstrous even if compared to our closest competitor, the great ape, but our truly unique property is the physiological (as opposed to habitual) upright walk which includes special anatomical features such as a narrow pelvis, the slightly inward-directed thigh bones, and plantigrade feet.[2,3] Upright-walking species show an occipital foramen at the base of the skull rather than the posterior aspect. These are unmistakable markers that are readily recognized in the fossilized form. Anthropologists will always be looking for ancestors that do not show these features because the old school says that the upright walk developed slowly from a quadruped animal by gene duplication and mutations. That proposal is of beguiling simplicity if one forgets the biology and biochemistry required for that transition. Looking back in time one is faced with the fact that the brain had all but disappeared at Lucy's (*A. afarensis*) developmental stage, yet she was fully bipedal.[4] To find the presumed quadruped ancestor would lead to insurmountable identification problems unless, as shown in a cartoon by an unknown artist, step-less melting of one form into the other can be observed.

Large brains relative to body-size are always associated with bipedality whereas bipedality is not invariably associated with large brains. Once a chimpanzee-brained creature has the features of upright walk it becomes a hominid, and when that hominid acquires cranial features that suggest a larger brain and when artifacts are found next to the fossils, it becomes a member of our genus *Homo*. Did the apes change, as in the old model, or was a new species introduced, as the Genomic Potential Hypothesis predicts?

The experts think of slow transitions but their own evidence says replacements. The analysis of a nearly complete skeleton found in the Afar Valley of Ethiopia told Johanson and his colleague that *A. afarensis* was a true upright walker with a brain about as large as that of a chimpanzee.[4] More than 10 years after this sensational

*The Genomic Potential Hypothesis: A Chemist's View of the Origins, Evolution and Unfolding of Life*, by Christian Schwabe. 

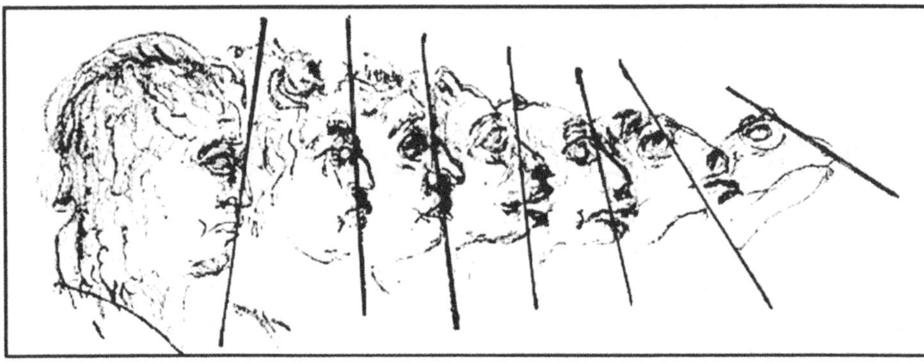

Fig. 12.1.Fig 12.1 . In this figure the requirements for establishing a smooth transition between species is depicted. While this example from an unknown artist is meant to amuse the viewer, it is precisely what Darwinian language projects when he speaks about the " inconceivably large number of intermediate forms " but he concludes — " if this theory be true, those have lived upon the earth."

discovery that destroyed the "brain first" idea, Johanson and his colleagues returned to the fossil hunt, this time in the Olduvai Gorge. He found a specimen that is, in my opinion, equally sensational because the creature was as small as Lucy, but lived about 1.8 million years ago compared to Lucy's ~3.3 million years.[5] The skeleton had very similar features, but the brain case was large enough to earn its owner membership in the homo family as *H. habilis*. There was a gap of 1.5 million years between *A. afarensis* and *H. habilis* which seemed to have resulted in a slight increase in brain volume but, remarkably, no other changes. A quote from the source of that information may help to illustrate this point for all its importance.

"If body height in the human line did indeed increase gradually from *afarensis* to *erectus*, then by heights *Homo habilis* should have averaged somewhere between four and a half and five feet tall. Instead, we had found a habilis skeleton that appeared to have stood no taller in life than Lucy herself. Judging from the fragments we had of the Dik-dik Hill hominid, from the neck down she was practically Lucy's twin.".[6]

This result was baffling since an almost contemporary of *H. habilis*, the 1.6 million-year-old nearly complete skeleton of the 5' 4" tall "Turkana" boy, had been found who was a member of a species called *H. erectus*.[7] This species had modern body proportions and significant mental capacity as signified by the fine Acheulean tools found near the fossils.[5] A new, more advanced species had appeared from nowhere!? Johanson made the following revealing statement:

"We already knew that *Homo erectus*'brains were significantly bigger than those of *H. habilis*—and now it seemed evident that their bodies were qualitatively different too".[6]

He left the ends hanging, but the message rings clear: whenever it was possible to follow a species for any length of time, very limited changes would be observed until elsewhere, in about the same layers, a more advanced species showed up! In all cases has it been impossible to establish unique connections to the purported ancestor in the hypothetical line-up of hominids. That is precisely what the genomist predicts, i.e., species do not branch, they develop, each from a specific pro-form within a million years or so, rapidly enough that only the nearly final stage enters our fossil archives. Everyone in evolution knows this scenario and accepts it as a conditional Darwinian event, confident that the intermediate forms required for credibility will show up in due time.

By the same reasoning one cannot be sure that any particular bipedal creature, say *A. anamensis, A.afarensis, H. habilis*, or even *H. erectus* would be in *H. sapiens'* ancestral line. There is no set of intermediates to make that point and without these intermediates the Genomic Potential Hypothesis explanation should be considered. Accordingly, hominids came in waves from independent clones, rapidly proceeding through the hypothetical initial spurt stages until they were large enough to leave identifiable fossils, likely smaller than the ultimate form. Finally, starting less than a million years ago, the brainy clones went through their juvenile forms to appear as *H. neanderthalensis* and *H. sapiens*. And again we read about these remarkable overlaps such as the coexistence of *H. erectus* and modern humans on the island of Java,[8] the well-documented overlap of *H. sapiens neanderthalensis* and *H. sapiens sapiens* in the Middle East.[9] Overlaps *per se* do not exclude branching, but when the fossils in question are as recent as *A. afarensis, H. erectus, and H. habilis* or even more recent, *H. sapiens sapiens* and *H. sapiens neaderthalesis,* the actual branching event lies so close that one should easily find the fossils. The rewards of finding branching points would have been very great indeed and an intense search to reach that goal is in progress. The fact that anthropologists have been unsuccessful weighs heavily against the branching scenario.

This view brings us onto a collision course, although it provides a more plausible place for Johanson's Lucy and her daughter than the old model. It is considered more significant to find a human ancestor than merely another biped that hit the stage prematurely so that usually every effort is made to make that human connection. To the uninvolved observer the evidence suggests that *A. afarensis* (Lucy) lived from about 4 million years ago until 1.8 million years ago and might have developed into *H. habilis* before extinction. At that time *H. erectus* had grown to the size of modern humans and persisted as *H. erectus* until he too met both, a newcomer *H. sapiens* and extinction. This I believe to be the essence of the story as concerns evolution. Anthropologists are quite rough with each other about the details and we have no reason to get involved. The information that seems to be fairly secure is that intermediates between the various hominids have not been observed. The evidence that has accumulated until now suggests that the biped physique appeared abruptly about 5 million years ago in Africa and that upright walkers that came later also came with the potential for larger skulls.

The "out of Africa" idea of human evolution is dominant among anthropologists perhaps because most fossils were found there, because of the commitment to a single origin scenario, and finally because of cultural demands that all humans be of equal descent. In the final analysis these are not very forceful reasons. The African desert may be just more accessible and provide for better preservation of bones. The old paradigm, however, is so strong as to merely allow discussions concerning migratory routes when fossils are found elsewhere; multiple origins are almost never considered. For the genomist there are good reasons why humans should have become visible in many places within the same time frame. Nevertheless, the "African origin" arguments will be resolved when human remains of such age are found in, say China or Australia[10, 11] that only air transport could have brought them from Africa in time to show up in the same horizons. As one looks into earlier layers the origin of these hominids should be discovered. The fossil record of other species reaches up to 100 times further into the past so one might reasonably expect to see the predecessors of bipeds at least into the Eocene (~50 million years) together with the bones of horses? But we do not! Bipedal creatures disappear from our view not by melting into the contours of a different animal but rather by fading out, i.e., by getting smaller rapidly and then disappearing. The genomic potential model suggests that the forms melt into a specific pool of primary stem cells and that these stages will be recognized once a mental image has been formed as to what to expect. Note that germinal forms of Cambrian animals, which the new hypothesis predicts to exist, may well have been seen already.[12,13]

Nobody has intentionally looked for germinal forms of large animals or humans whereas everybody has looked for intermediates. A straightforward reading of the evidence produced by over 100 years worth of anthropologists' efforts to find them has failed. The selection of acceptable interpretations is very limited when a species, clearly marked in the fossil record, suddenly disappears as one proceeds into lower horizons while other animal fossils persist undiminished in the same layers. Either the species emerged at once from parents of a very different kind (the "motley egg" hypothesis), or it did what species had to do when there was no macroscopic ancestors around, i.e., emerge from a species-specific primary "stem cell" that had developed during the Precambrian. This is the metamorphic event postulated by the Genomic Potential Hypothesis which stresses credulity because it is counter-intuitive. A 'caterpillar' in the ancestry of all animals, including humans, is a tall order even for the most propitious of readers, so let me take time out to bring the problem into perspective because it is as crucial for human evolution as it is for all other species.

All eggs and seeds are sensu stricto metamorphic stages. Birds, amphibians, reptiles, and even some mammals have external eggs from which, after a period of incubation (development) the juvenile form of the species emanates with just the cracking of a shell. Insects go through this stage as well but, in this case, the first stage is another "egg", a caterpillar, a moving, feeding egg that increases the energy reserves and then wraps itself into a cocoon from which in due time the amazing color stencils of butterfly wings emerge. Two completely different body plans, different locomotion, and different feeding patterns produced not slowly by mutations and adaptations, but read from the genome in abrupt stages from one start. Had this mode of propagation gone extinct about 10,000 years ago, and even if we would be able to see butterflies preserved in a fine-grained shale, the connection to caterpillars and pupae would never have come to light without a hypothesis to that effect. Experts, driven by the old model, would keep looking for the oldest possible butterfly and beyond that, for a fly or a flightless bug, as they actually do,[14] that could have been converted by sunlight and an avalanche of gene duplications and mutations into a butterfly.

Secondary ova, those that were produced by a finished prehistoric animal, have been found frequently, and when dinosaur eggs revealed the fully formed embryo there was no doubt about the saurian way of reproduction.[15] Precambrian eggs have been found[16] but never one of the primary eggs, which must have been there en mass for each species a few thousand years before the entry of a species into the fossil record. Until now, there was no good reason to look for them and, of course, there is no handbook for identification purposes. Certainly the avalanche of exquisite fossils from China, which is pushing the age of vertebrates deeper into the Cambrian, will provide an excellent chance to find some of the pro-forms postulated by the Genomic Potential Hypothesis.[17]

When did these primary ova give rise to the first metazoan with human potential? The fossil record suggests that it might have been between 5 and 3 million years ago. It is true that several clones of primate quadrupeds (early ape-like creatures) began to appear about 15 million years ago. They have colorful names like *Ramapithicus africanus, Sivapithicus africanus, Kenyapithicus africanus, Kenyapithicus werteri* and Proconsul *africanus,* to name a few. Within the logical framework of the Genomic Potential Hypothesis it seems that this group represents an isolated eruption of ape-like creatures, which do not connect to modern apes or humans other than to foreshadow the arrival of higher primates. This view appears to be shared by at least some anthropologists.[18] The early record of nonhuman primates seems to be interrupted for a period beginning about 8 and ending 5 million years ago when modern ape fossils left their marks slightly ahead of the first hominids (upright walkers, 4.5 to 3.5 million years ago). This was the time at which *A. afarensis* (Lucy) lived who literally rose from the dust in 1974 with significant help from Johanson and White.[4] While hominid skulls and bones had been found by other anthropologists, the Leakey family comes to mind, which were named Australopithicine*s* (southern apes), no leg bones or pelves of such age had been seen from one individual before Lucy.[19] Lucy came as a shock to anthropologists because her knee bones, her feet, and pelvis showed clearly a fully finished upright walker, but her skull was as small as that of a chimpanzee. Lucy was not terribly popular among students of evolution, the "brain-first" paradigm lain in shambles at her feet (she would have known that it is better to be infamous than ignored). It was Lucy's "hypothe-cidal" quality that drove the fiercely competitive and secretive stars among fossil hunters back to the fossiliferous grounds in Africa to see what came before her. As a consequence Lucy got an ancestor called *A. anamensis,*[20] who was bipedal and lived between 3.9 and 4.2 million years ago. More appear to be rising from the parched landscape of Ethiopia that could extend bipedality to 4.4 million years when the official entry for *A. ramidus* is made.[21] That is where we are, skin deep into the history of life. The large skull has long been lost on our travels into the past but the upright gait persists. This reality points to another conceptual difficulty inherent in the search for transitional forms that lead from animal X to hominids. Once the brain is lost as well as the upright gait there remains no identifier that would tell us that the suspected ancestor is not a member of one of an extinct species of prosimians, for example. A speciation fork is unprovable unless two paleontologists, each following a different fossil into the past, will meet at the same fossils in a lower formation to claim it as the ancestor for his species. The new hypothesis demands that fossils never meet so that the two observers, starting with different fossils, will never see each other again. The sketch below shows the world lines of early and recent primates dangling into the past, disconnected from the pools of Darwinian ancestors and reattached to the primary ova of the Genomic Potential Hypothesis.

There is no sign of a transition from quadruped to biped or from ape brain to the larger brains of hominids. Fossils of small and large brain cases show up side by side until one group disappears while the other continues the relay.

One might ask, for instance, what happened to the apes during the past 5 million years? Nothing changed in any remarkable way other than an increase in size. They could have developed brain, but it was not in their genes. The conditions for Darwinian selective pressure were definitely there. A quadruped with a larger brain would have done substantially better than one with an ape brain. It would have helped them to avoid a future life in zoos and the cages of their experimenting "cousins". Victims of

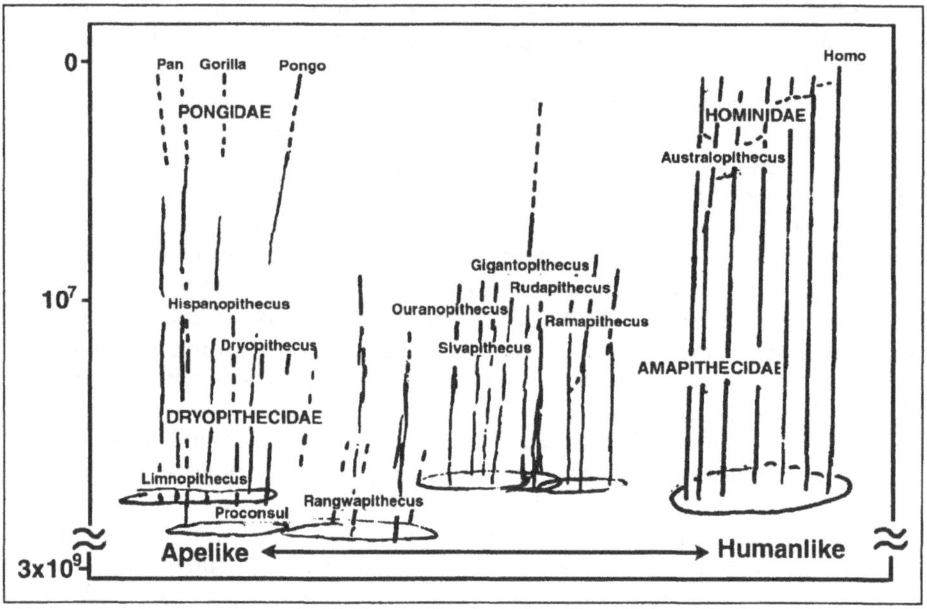


Fig. 12.2. This is a take-off on a figure by paleontologists concerning the evolution of nonhuman primates and hominids. The worldlines of species from the Genomic Potential Hypothesis were added to the figure. Beginning in a circular region at the bottom which depicts a clonal relationship during primordial times, the species rise to the present or to extinction as relatives from a chemical pool as opposed to branches of a tree.

an unmodifiable genome, modern apes merely watched how *A. afarensis* raised their young in that awkward upright position (equally without a choice). The robust Australopithicines in turn watched a larger-brained upright walker called *H. habilis* for a while, and possibly had him for dinner off and on, because *H. habilis* was somewhat smaller, before both headed for extinction (not adaptation). It must have been a strange experience for *H. erectus* to meet the first of *H. sapiens* as they apparently did on the island of Java and possibly elsewhere.[22]

The fact of the matter remains that nearly all of those elevated retroactively to the status of human ancestor lived together for part of the time of their existence. They materialized in our records from splinters a little larger than dust specks, one outline after another, brought into full view by the skills of anthropologists, as more lined up as outlines in the distance. But there is insufficient evidence to line them up in a bona fide chain of ancestors, they were most likely covivants on earth.

The main story leads us to two brain centers, both appeared between 500,000 and 200,000 years ago, one known as *Homo sapiens neanderthalensis* and the other as *Homo sapiens sapiens*. The brain was large in relation to body size with Neanderthales enjoying an edge of a few $cm^3$ over sapiens. From there on the development of more potent brains was a matter of selection and breeding, the size of brains remained fairly constant once *H. sapiens* was recognized in its earliest forms. Even today the brain power of individual humans varies by orders of magnitude, but the extremes of humans and apes do not nearly overlap in spite of what we at times contend about each other.

For the time being *A. afarensis* or possibly *A. anamensis* are the first upright walkers, our preludes sensu lato. Questions seeded by the Genomic Potential Hypothesis pertain to the path between any of the first bipeds and their corresponding primary stem cells. How would this transformation look in a real world that had predators lurking behind every bit of green? Could one expect a millimeter size upright walker to exist and to fend for itself at a time when every legitimate grub or insect would be elephantine by comparison?

It is likely that they went through survival stages, moving, feeding larvae, protozoan, fungi[23] or different durable creatures of that size and would not transform into much larger organisms until, at the end, a genetic upright walker would crawl away on hands and short legs from the last metamorphic event while the genomic potential would drive its size up to the point where it could assume the physiological upright position and remain alive in a competitive world. The same had happened at different times for *A. afarensis*, for the other Australopithicines, for *H. habilis, H. erectus* and so on, for many groups that we have yet to discover. If this thought is uncomfortable it is so only because we did not grow up with it and because our self-esteem has provided for a more grandiose scenario.

One must view the new proposal against what has been accepted so far. Many, if not most, Darwinians are often unaware of the details of their model. Here is a short version of it.

Looking back from our branch in the traditional tree of life, right at our feet, we see several lines of hominids and, close by, the great apes. Further down lemurs and tree shrews may have carried our genomic torch, but as the deep Tertiary is left behind the choices become limited. A nondescript mammal in the form of a small saurian comes to mind, a hairy reptile has been mentioned in the Cretaceous and perhaps a reptile in the Jurassic. In the Devonian the Darwinian life becomes waterlogged, leaving us fishes, brachiopod-type clams, or worms like priapulids as potential banner bearers for the pro-human genome. This would get us through the Ordovician and the Cambrian and perhaps into a Precambrian where sponges were found in the 650 million year old phosphate deposits in eastern China and central Russia.[24] Now, we are passing through 3 billion years of single cellularity, until we end up within the cubic micron confines of the single cell that started it all. This conjures up images of the one microscopic cell on this earth as seen from an airplane window, and that is the point where the old idea moves into the proper perspective. On this time-reversed trip to this cell we crossed kingdoms, phyla, classes, families, and species, with only one recurrent explanation, i.e., random mutations. It is not the nightmarishly difficult concept that makes this model impossible but rather the lack of contiguity that enters the paradigm with every "constructive" mutation (see Chapter 2).

The precedence argument may help to clarify, if not resolve the issue of evolutionary routes. The Cambrian has given the stamp of reality to the transition from single- to multicellularity. There were only calcareous microfossils in the antecedent Tommotian, but there were masses of arthropods in the mid-Cambrian! That would mean that different, to us invisible, life forms had undergone transformation into arthropods! Thousands of transformations, one for each member of each species, had occurred according to the Genomic Potential Hypothesis. In the Darwinian model it was one for all metazoan, followed by descent with variations, a mechanism for which the time was too short. We have visited this scene before (Chapter 7); it is a crucial one and it

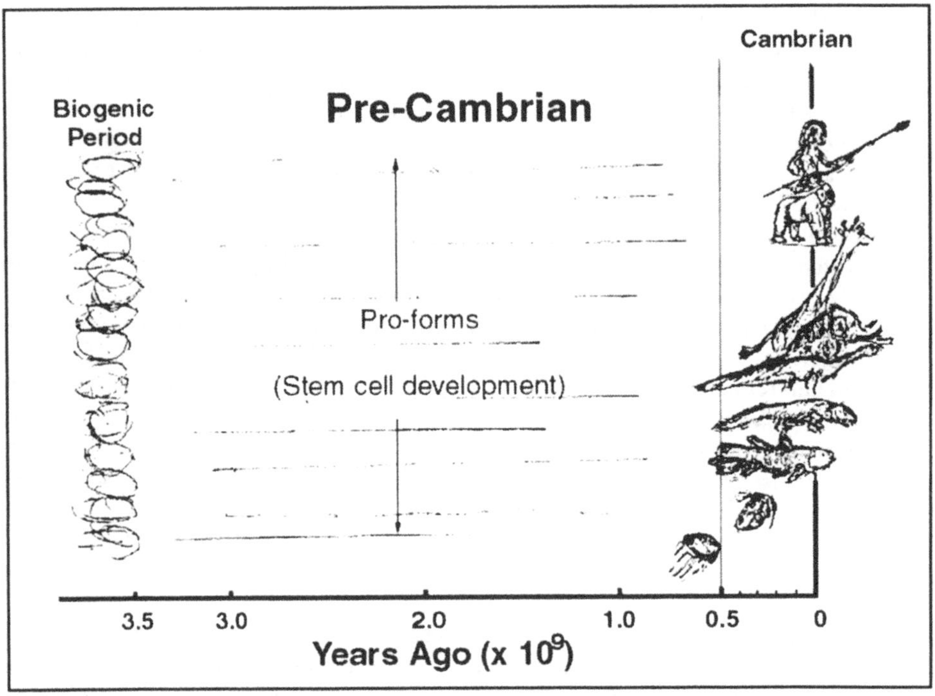


Fig. 12.3. This drawing transmits a realistic concept of the time of development of metazoa as a function of the total biogenic time on earth. Notice how short the distance is between a jellyfish and a human and how little time there was in the old model to convert any of these established species into a completely different body plan. The Genomic Potential Hypothesis gives each species an almost equally long time of development during the Archean period.

tells us that the current model of evolution is dogmatic in the face of contrary evidence. The new model predicts the simultaneous appearance of groups of animals in successive intervals such that one does not need to invent new mechanisms for every animal; it is the same type of transition occurring a little sooner and a little later (by geological standards).

While the new hypothesis depends upon the observations that intermediates between species are missing, the uncertainty about our germinal stages remains until one finds the metamorphic state frozen at the moment of conversion to a phenotype in the late Quaternary sediments. Science, at our present state of development, leaves us no alternative.

As surely as intermediates between species are nonexistent, the intermediates between the primary ovum and the first stable fossilized creature are there for us to discover. They will be in the same horizon, only a few thousand years or less apart. Figure 12.3 gives the impression that all species had been washed into the present by a powerful wave so that the times of arrival are only minutely different. The time axis gives one a realistic impression of the awesome speed of appearance of species when compared to the total time of life on earth. The whole colorful world of anthropology that reverberates from fierce battles over what is a common ancestor, what limb should be attached where on which evolutionary tree, all of that fits into a little more than the width

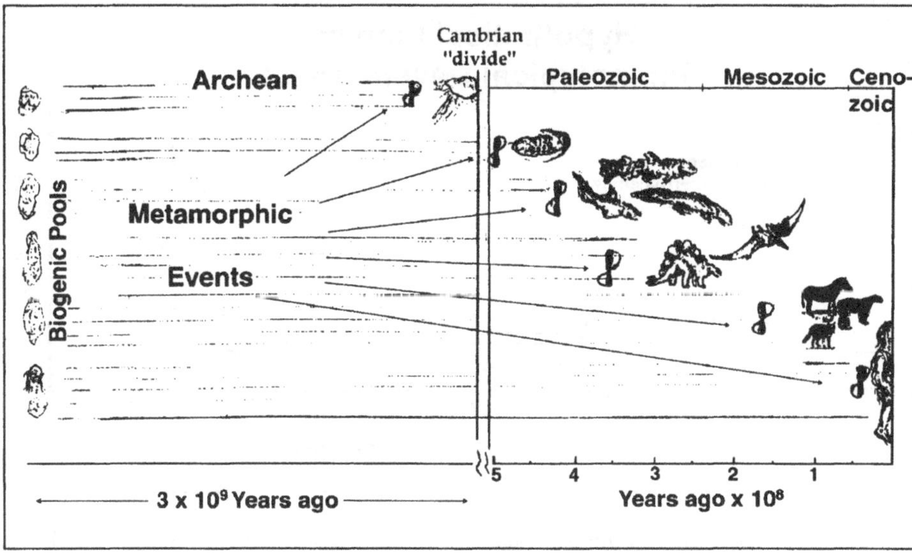


Fig. 12.4. In this figure the time of metazoan development has been expanded in relation to the total biogenic period in order to provide space to show where the metamorphic processes might have occurred which, followed by a rapid spurt of development, led to groups of somewhat similar species that we group today as order, suborder, genus and species.

of the line frame that separates past and present in Fig. 12.3. To an uninvolved observer the intensity of the dispute about the common ancestor means that none has been found. The Darwinian hypothesis pours fuel into the flames by insisting that there is a common ancestor and if they look long enough they will find it. The new hypothesis is kind to its followers, it provides a unique first ancestor for each of the hominids.

In the next figure the scale has been expanded to show how the transition to phenotypes can be anchored to indisputable events which, together with the absence of intermediates between animals, forms the foundation for major predictions derived from the Genomic Potential Hypothesis. Reference is made to the precipitous appearance of arthropods during the mid-Cambrian period right on the edge of the microfossil-producing Tommotian age. There is no identifiable macro-organismic precursor so that here the Darwinian chain of descent with variation is forcefully interrupted. The process must be continuous, however, and the only explanation left is metamorphosis. The implications are presented in Fig. 12.4. The 'violin keys' designate the approximate time when the conversion of stem cell clones began to produce 'metamorphs' in large numbers within a period of time of which many persisted to fill the ranks of families, superfamilies, a process that may be a target for future research.

The Cambrian arthropods start the series and because, as mentioned above, there were no macroscopic ancestors that could have been converted slowly by mutation and adaptation to the first animals in the shale, they set the tone for the evolutionary calendar of events. There are no recognizable intermediates leading to fishes, would it not follow that they too should have evolved by the same route that brought about arthropods? The step from fishes to amphibians is informative. *Coelacanth* and its purported successor *Ichthyostega* are contemporaries. Would that not suggest quite

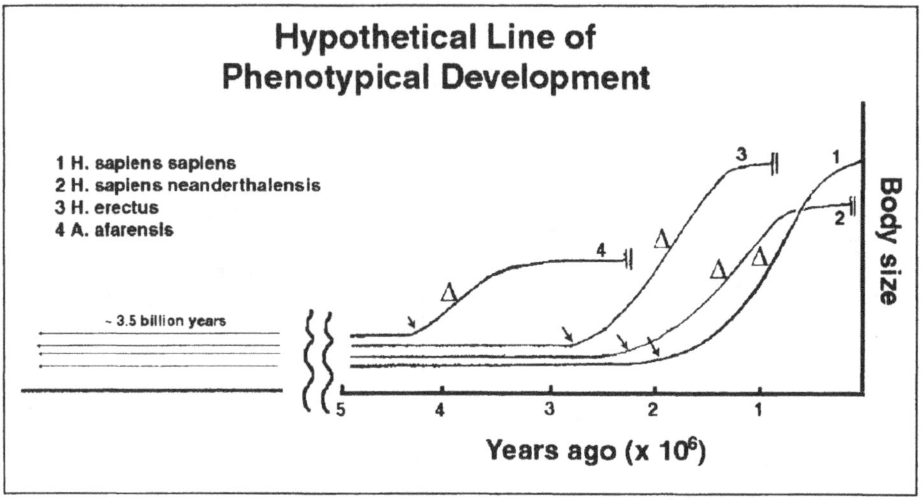


Fig. 12.5. This gives an even more expanded view of the last 5,000,000 years, the time of homonids. At this scale one sees distinct differences, for example, between the origin of *A. afarensis* and the time when it became visible (arrow points) until extinction. Notice that *H. erectus* origin is somewhat later but the species is well established before the extinction of *A. afarensis*. Again, *H. neanderthalenses* and *H. sapiens* overlap *H. erectus* and probably numerous other homonids yet to be discovered.

forcefully that both came about by the same mechanism? And the mammals a little further along and so on. The answer must be the same for all these cases until humans came along, and there is no evidence that would compel us to use a different explanation for our evolution. This model again leads to predictions as discussed in the context of the next figure where the scale has been extended to make our history visible.

The arrows in Figure 12.5 show the time of the purported metamorphic event and the triangles show the point during development when the species should be anatomically distinct, albeit smaller than the final form. The scale for hominid evolution has been expanded such as to include only the last 5 million years, i.e., 0.15% of the total biological eon on earth! Most readers will find the implications of this figure deeply disturbing because our place in the evolutionary tree had become part of our culture. The view from the tree top was great and besides that, we have lost all our "inferiors". But how could one exempt humans, by what contiguous scheme could we make ourselves different? All autonomic human functions are not better (sometimes less efficient) than in an animal. Of course, we are outstanding as concerns the central nervous system but it is not unique, it is merely more of one thing and perhaps differently folded. Within the framework of the Genomic Potential Hypothesis we are fitting perfectly well into the scheme of things.

This is the point where the hypothesizer clears the field for the experimenter. If the series 'caterpillar-pupa-moth' would have to be extracted from the fossil record perhaps we could get the evidence today because we know that it happened as surely as we know that the caterpillar did not go through descent with variations to become a butterfly. Perhaps with a new attitude we may be able to find our developmental pro-form in the lower Quaternary layers. Metamorphosis teaches that the gene is important for what an animal looks like, and when one can accept this lesson then one can make peace with a fossil record that does not want to produce the intermediates that everybody is looking for.

There is our edifice of concepts, beautiful and a little aloof. Looking from the balcony into the street we can see anthropologists still arguing that Lucy was not quite the upright striding creature that the discoverers made her out to be, but that her gait still had a twist of "intermediate " quality to it. Then we see Mary Leakey with the Laetolil footprints of *A. afarensis*' contemporaries, which marked the ground, measure for measure the way we would, and even then the old hypothesis would not give up. What is the point? Lucy killed a bad idea, which is commendable; she never changed into anything else. She was likely not our ancestor but rather one of the group of upright-walkers that ran parallel and a little ahead of *H. sapiens*. She shared with us the gait but not the brain. One that shared both, the Neanderthal, came a little ahead of us. The last 150,000 years we lived together, trading and alternately occupying caves in the Middle East[22] and recently even sharing the front cover of the Scientific American under the title "Once we were not alone".[9] What makes me chuckle at the sight of such politically correct display is the fact that between existing branches of *H. sapiens* (to which Neanderthal belonged) the artist could have found significantly larger differences than those he chose (dared) to display. So why are we alone? Because we say so, that's all. In our enlightened age we cannot separate science and politics. The reader may remember my introductory remarks about the amalgamation of science and culture and correctness.

To do science properly one has to be willing to be politically incorrect. This building of mine is politically incorrect (apolitical in fact), but it is esthetic and elevating to the free spirit, and some day evolutionists will look into it and suffer an attack of deja vu and then our theoretical excursions will acquire a different glow and likely a new owner to boot. Chemistry is Machiavellian.

Darwin is often accused of having never dealt with the origin of species other than in the title of his book. The Genomic Potential Hypothesis tells why he could not have come any closer then he did, and history will recognize that he picked the only nondivine choice left under the circumstances, the breeding of one species from another by infinitely extended animal husbandry. Science was not ready to provide the limits for that approach; science was not ready for Darwin.

## References

1. Gould SJ. The persistence of natural philosophy. Evolution and the triumph of homology, or why history matters. Am Scient 1986; 74:59.
2. Lovejoy CO. Evolution of human walking. Sci Am 1988:118.
3. Crelin ES. The development of the human foot as a resume of its evolution. Foot Ank 1983; 3:305.
4. Johanson D, Edey M. Lucy. New York: Simon and Schuster, 1981.
5. Johanson DC, Masao FT, Eck GG et al. New partial skeleton of Homo habilis from Olduvai Gorge, Tanzania. Nature 1987; 327:205.
6. Johanson D, Shreeve J. Lucy's Child. New York: William Morrow & Co., Inc., 1989.
7. Brown F, Harris J, Leakey R et al. Early Homo erectus skeleton from west Lake Turkana, Kenya. Nature 1985; 316:788.
8. Swisher III CC, Curtis GH, Getty JAG et al. Age of the earliest known hominids in Java, Indonesia. Science 1994; 263:1118.

9. Tattersall I. Once we were not alone. Sci Am January 2000:56-62.
10. Thorne AG, Wolpoff MH. The multiregional evolution of humans. Sci Am 1992:76.
11. Wolpoff MH, Hawks J, Frayer DW et al. Modern human ancestry at the peripheries: A test of the replacement theory. Science 2001; 291:293.
12. Pratt BR, Zhang X-G. Unscrambling fossilized Cambrian embryos. J NIH Res 1995; 7:46.
13. Bengtson S, Zhao Y. Fossilized metazoan embryos from the earliest Cambrian. Science 1997; 277:1645.
14. Gould SJ. Bully for Brontosaurus. New York: W. W. Norton & Co., 1991.
15. Morell V. Follilized hatchling heats up the bird-dinosaur debate. Science 1997; 276:1501.
16. Chen J-Y, Oliveri P, Li C-W et al. Precambrian animal diversity: Putative phosphatized embryos from Doushantuo Formation of China. Proc Natl Acad Sci USA 2000; 97:4457.
17. Stokstad E. Exquisite Chinese fossils add new pages to book of life. Science 2001; 291:232.
18. Pilbeam D. The descent of hominoids and hominids. Sci Am 1984:84.
19. L.S.B.Leakey, P.V.Tobias., J.R.Napier. A new species of the genus homo from Olduvai Gorge. Nature 1964; 202:7.
20. Leakey MG, Feibel CS, Mcdougall I et al. New specimens and conformation of an early age for *Australopithecus anamensis*. Nature 1998; 393:62.
21. White TD, Suva G, Asfaw B. *Australopithecus ramidus*, a new species of early hominid from Aramis, Ethiopia. Nature 1994; 375:306.
22. Swisher III CI et al. Latest *Homo erectus* of Java: Potential contemporaneity with *Homo sapiens* in Southeast Asia. Science 1996; 274:1870.
23. Redecker D, Kodner R, Graham LE. Glomalean fungi from the Ordovician. Science 2000; 289:1920.
24. Li C-W, Chen Y-J, Hua T-E. Precambrian sponges with cellular structures. Science 272; 279:879.

# Chapter 13

# Molecular Genealogy

Is it rational to expect proteins such as insulin, relaxin, hemoglobin, or cytochrome to have followed the evolutionary route of the organisms from which they were taken? Of course it is , but it does not follow a priori that protein structures would reveal relatedness between species.[1, 2] The Genomic Potential Hypothesis predicts no branching of either species or proteins and since proteins within species are the same there is harmony between proteins, species, and reality. In contrast, the old model starts with a single origin and produces variety by continuous branching (changing genomes and hence proteins under continuous mutational pressure). While the proteins within species should still be alike, the proteins from neighboring species must show sequence differences proportional to the phylogenetic distance deduced from the fossils. Of course, the purpose of molecular genealogies is to deduce species' relations from proteins and it is easy to imagine the turmoil created when proteins, such as pig relaxin, project a pattern of relatedness that is totally incompatible with the old branching model. Since an animal and its proteins are going through evolution (by any model) together, all proteins must give the same pattern of branching or no branching at all. Failure to meet this criterion means that no part of that hypothesis can survive.

To test for a branching order, a strong condition has to be placed upon the model, which reads that "protein sequences reveal relatedness among species". A less stringent condition such as "some protein sequences reveal relatedness among species" is useless because it requires another parameter to determine which protein should be used. In spite of substantial reservations, the concept of molecular genealogy has achieved a superior status in matters of relatedness even among paleontologists. The idea took root with the discovery by Pauling and Zuckerkandl[3] of hemoglobin and cytochrome structures, and the first message that seemed to crystallize from the data was that species of recent vintage had similar proteins and that species with a long history had proportionately modified ones.[4, 5] A simple sketch may depict the basic logic of the neo-Darwinian molecular genealogy as it relates to their speciation model (Fig. 13.1).

The meandering line depicts diversification and genetic mixing of genomic material within species. The first life on earth starts with one organism, expands rapidly, and keeps the genomes uniform by interbreeding. An obstacle separates the group so that the mixing action now excludes half of the genetic reservoir. Logic tells us that both groups, each now limited to a fraction of the original genomic pool, would change under a regimen of random modification. Thus, the subgroups would change their characteristic at every one of the obstacles, creating more species while erasing the parent species from the fossil record. Such a branching scenario would limit the

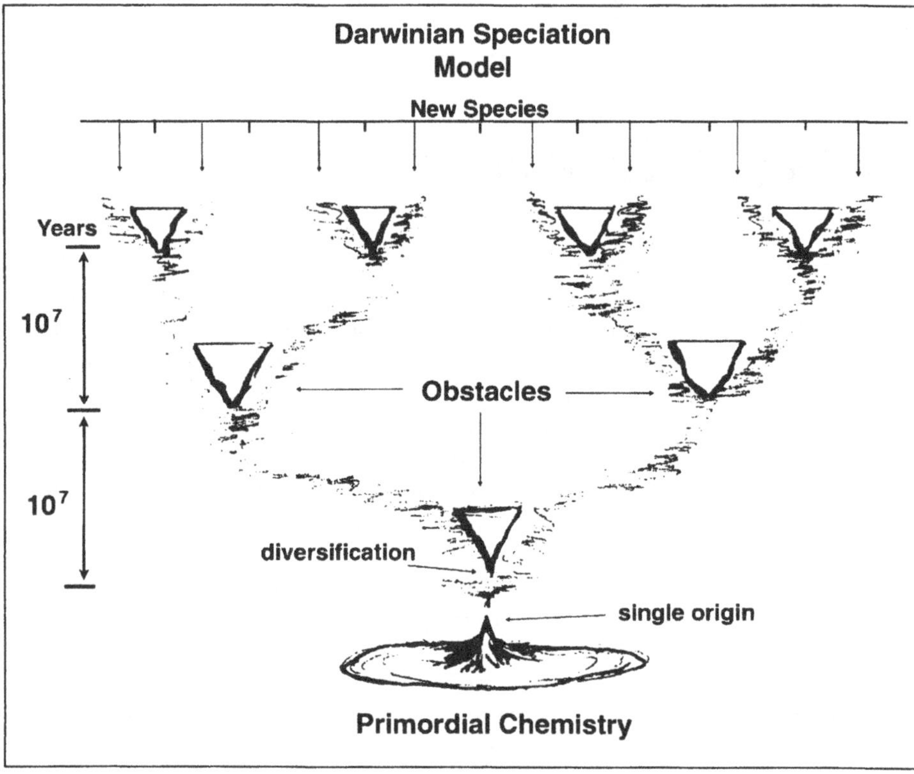


Fig. 13.1. This figure illustrates the Darwinian model of speciation which starts from the single origin producing one clone that is separated by an obstacle into two groups which subsequently accumulate random mutations so that they become sufficiently different from each other that they could no longer interbreed if they were brought together again. This scenario repeats any number of times, splitting the original clone into all of the fauna and flora that we observed today. The critical point brought out by this picture is that once a species is divided, both branches should change to an equal extent if random processes are involved. Note that the Darwinian hypothesizers have overlooked that fact.

persistence of any species to about 10 million years whereas evolutionists of any persuasion will agree that species last much longer. This logic did not register in the Darwinian camp where branching is viewed as an event that leaves the original species unchanged and, without explanation as to the mechanism, lets only the new buds mutate toward a new identity.[6] The same error is incorporated into the neo-Darwinian scheme of molecular evolution by gene duplication and mutation where only the new gene develops a different function whereas the old gene is protected from mutations because it encodes a vital protein. It so happens that random processes are neither selective nor influenced by special needs and would hit both copies of the gene with the same average frequency.

Since mutations change the genome, the argument continues, the mutations that caused speciation continue to produce changes in proteins, and because mutations do not stop after this event, the percent sequence difference observed in parent and daughter species would be proportional to the time that has passed since separation occurred. Species produced at the first obstacle in Figure 13.1 would show larger differences between homologous proteins (insulin for example) than those produced at later segregations since all have continued to live to the present[4] (proteins can be recovered only from living species). This thinking led to the concept of an evolutionary clock that ticks off time on a dial of protein sequences. Every clock needs to be standardized according to what is to be measured, which in this case would be geological time as a function of a chemical parameter, i.e., protein sequence differences. For that purpose one would need two species with known fossil records and determine the sequence difference of their insulins. One would ask how fast has insulin in pigs mutated during the 50 million years by comparison with human insulin? The difference is taken as a mutation rate, i.e., number of mutations accepted per unit time. Porcine insulin has about 50 amino acids and differs by one residue (2%) from human insulin. Differences used to be expressed in Pam units which are defined as point mutations accepted per 100 million years per 100 residues or percent change per 100 million years. The fossil record of pigs is about 50 million years which is taken as the branching date. This would amount to 4% in 100 million years or 4 Pam units. Accordingly, insulin mutates at the speed of 4 Pam units which then becomes the constant whereby all other relations are established. Thus, every 2% difference means 50 million years ago as concerns branching time. Human and bovine insulin differ by 6% and hence cows and humans had a common ancestor about 150 million years ago. Pigs and guinea pigs differ by 33% so that the last common ancestor of them lived some 800 million years ago.[15] The fossil history of guinea pigs, however, is contemporary with that of histricomorphs (porcupine), mustelidae and muridae, i.e., about 50 million years or less. The muridae do have very active insulin so that the inferior insulin must have selectively spread through the whole guinea pig population after branching. The argument brought forth at this occasion is that insulin in guinea pigs was good at first but it mutated faster than in other species. During a meeting in Florence in 1988 I had a serious argument on the floor, questioning the speaker as to how he would justify a poor insulin (which it is) to have spread though the whole guinea pig population when selection is guided by improvement? "Ah", he said, "but this insulin has a second function in the guinea pig that is as yet unknown!" This was not anybody talking gibberish, it was a well known protein crystallographer. All constraints have fallen; if homologous proteins in closely related animals are too different, one of them mutated faster and, if such proteins in distant species are identical, it is due to lateral gene transfer. What a pleasure it must be to be an ordained member of that cult. This is one of the fundamental flaws that will eliminate the old hypothesis of evolution with or without my help.

The late Allan Wilson[7, 8] had made himself quite unpopular with his colleagues by proposing that proteins would accept mutations at a uniform rate. For molecular phylogeny to be meaningful, Wilson's molecular clock had to be right in as much as he described precisely what was implicit in the paradigm. If proteins were to stop and to speed up their mutation process at various times, one would need (unknowable) independent parameters to take these rate changes into consideration. From the guinea pig data and many other examples, one can surmise, however, that Wilson's critics prevailed and it seems like fool's play that they celebrated their victory, not realizing that they had observed the demise of the whole concept of molecular genealogy and had, in the process, severely shaken the idea of descent with variations.

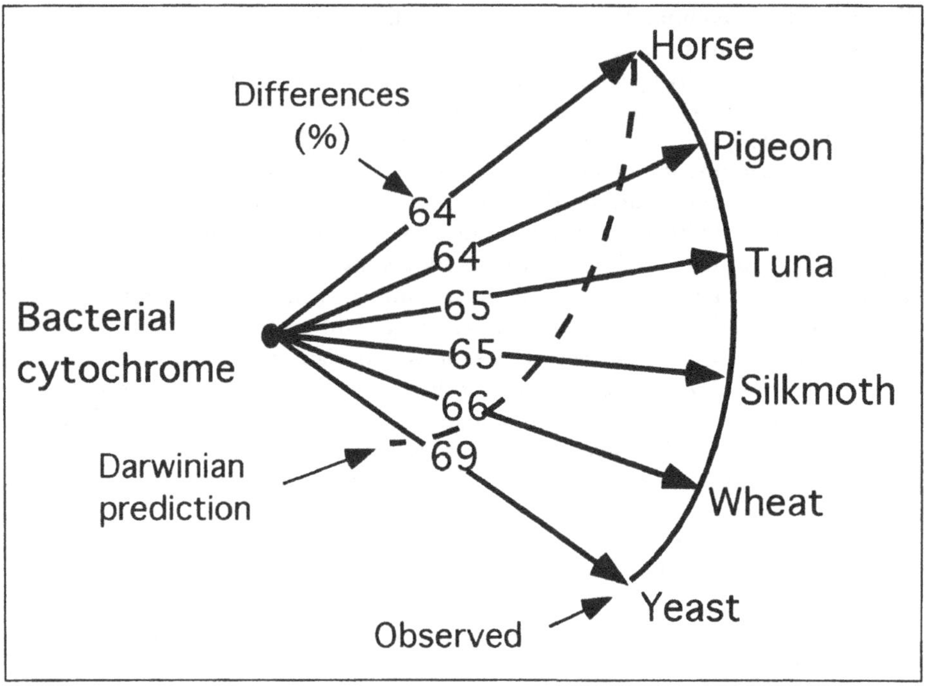


Fig. 13.2. This figure illustrates a lack of progression between the various taxa from single cellular organisms to man. If one form of life would have come about as depicted in the previous view, those that split early should be much more related to each other in terms of their protein structures than those that separated during the last few millions years, like reptiles and mammoths, for example. The dotted line in this graph would represent a Darwinian prediction whereas the solid line represents observation which is in harmony with the genomic potential hypothesis.

True, the most recent entries into the fossil record all seemed to have more similar proteins and so did the other groups roughly in the order of appearance of fossils. But there was no connection between the layers.[9] If one uses bacterial cytochrome as a basis of comparison, it becomes quite obvious that there is no continuous gradient of relatedness from yeast to mammals. No progression could be observed that would be expected if every new taxon is the gene duplication product of the previous one. The bacterial molecule is as distant from the yeast, its closest relative, as it is from a mammal and every species in between (Fig 13.2). Notice that the old model should have given rise to a relation as indicated by the dotted line. The percent sequence difference, which is 65% ± 5%, should have gone from near zero difference between bacteria and yeast to 69% difference between bacteria and a mammal. Within each taxon such as mammals, birds, fishes, insects, plants and fungi, the sequences of cytochromes is much more uniform, leaving open the possibility that these groups developed in clones.

Cytochrome and insulin are not unique in their opposition to the Darwinian model of genealogy. Relaxin, a hormone of parturition in placental mammals, shows clearly that molecular structures and branching patterns do not connect. When the sequence data of all known relaxins are reviewed, several startling observations could be made. The hormone differs by about 55% in animals of purportedly

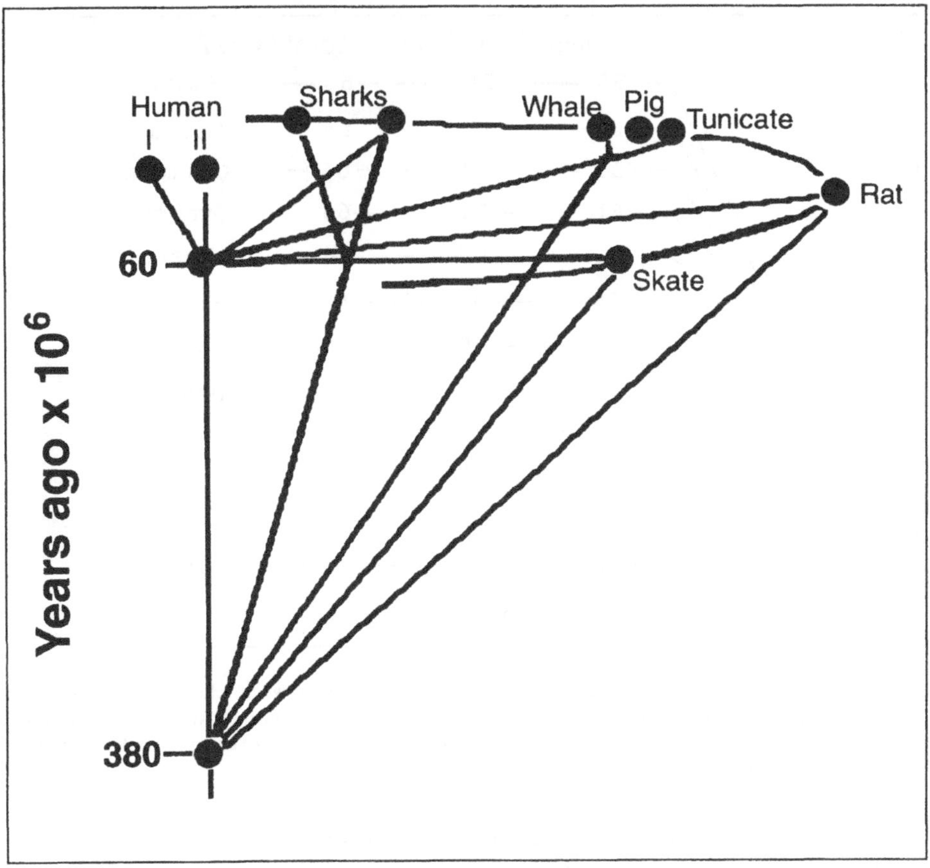


Fig. 13.3. This figure provides another example of the incongruity of fossil records and molecular genealogy. Using humans as a standard, all mammals that purportedly are closely related have relaxins that are not more closely related to human relaxin than to that of skates and sharks. Since skates and sharks have been around in the Devonian period 380,000,000 years ago, the structural dissimilarity of the various relaxin molecules would suggest that the branching point between humans, horses, rats, etc., as well as sharks would have occurred 380,000,000 years ago. Conversely, the mammalian explosion is placed at about 60,000,000 years, which would force all the molecular genealogy lines to converge at a point 60,000,000 years ago, giving us a 300,000,000 years uncertainty. This difference is not trivial, it represents 3/5 of the total time of macro-organismic development.

recent divergence and by about the same amount from the relaxin of chondrichtians (sharks, rays).[10-12] By the rules of molecular evolution this would indicate that chondrichtians and mammals diverged at the same time from the main stem of life either 60- or at 380 million years ago, i.e., the fossil record puts a 300 million-year distance between them. From this set of observations one can derive a choice of trees[13] with branching points either in the Devonian or in the Tertiary (Fig. 13.3). The difference is not trivial since 300 million years represents three-fifths of the total time Darwinians have allotted to metazoan evolution.

What is one to make of it? The cytochromes within classes are fairly similar and that observation has given rise to the idea of succession. As Denton[9] pointed out, the

Skate

| | B chain | A chain | total (x 1/2) |
|---|---|---|---|
| Human II | 55 | 71 | 63 |
| Gorilla II | 61 | 71 | 66 |
| Chimpanzee II | 57 | 71 | 64 |
| Orangutan II | 61 | 71 | 66 |
| Rhesus monkey | 61 | 67 | 64 |
| Pig | 71 | 67 | 69 |
| Rat | 71 | 79 | 75 |
| Dog | 68 | 75 | 71 |
| Guinea pig | 68 | 71 | 70 |
| Whale | 68 | 67 | 67 |
| Shark | 43 | 71 | 57 |
| Silkworm | 75 | 71 | 73 |

(percent difference)

Fig. 13.4. This figure reinforces the picture of nondiscontiguity between relaxin structures and the purported age of the owners of these relaxins. Skate is no closer to humans than to a shark or to a silkworm in terms of the molecular genealogy record whereas the skate, of course, lived 300,000,000 years before humans and other mammals appeared on earth.

all-important continuity is missing that would make these layers into steps of an evolutionary stair. The lack of continuity is equally obvious in the spread of relaxin sequences in Fig 13.4.

Relaxin does not even show uniformity among land vertebrates but one of them, the pig, is nearly identical to whales which as a group appear to have identical relaxins (two baleen whales and one porpoise). This would make the pig a bridgehead of sea mammals on land.[14] Before proposing new hypotheses please consider the recent discovery of a relaxin gene in tunicate gonads by Dr. Danielle Georges (UFR de Biologie-Université Joseph Fourier-Grenoble), and the partial sequence analysis of the isolated protein which left no doubt that tunicates have porcine relaxin! Actually the pig has tunicate relaxin since the record of tunicates dates back to 500 million years. Of course, the animals used are contemporary but, in as much as their appearance has not changed, it is a reasonable assumption that the proteins did not change either. The old idea of evolution by adaptation is unable to deal with this evidence. After this paper had appeared in the highly visible *FASEB Journal*,[15] the editor came under so much Darwinian pressure (under the pretense that contamination must have occurred) to retract or qualify our paper that he turned to me for comments which were printed as follows.

> Dear Editor:
>
> The observations presented in our paper suggest that sequence comparison is not a sure way to establish genealogies. This is a tall order and the justified questions proffered by some readers, although not unexpected, are sensustricto, unanswer-

> able. What is enough care? Dr. Georges has extracted gonadal and intestinal tissue from the tunicate and only the gonadal tissue contained the mRNA for relaxin. She tested ovarian tissues during the reproductive period and during the reproductively silent period, and only the ovarian tissue harvested during the spawning months produced mRNA for relaxin. These procedures have been repeated many times during several spawning seasons (about 3 years) with the same results. PCR errors would not be that consistent. Dr. Georges is an invertebrate biologist who has never worked with porcine material.
>
> The tunicate gonads that were harvested for protein extraction never saw my colleague's laboratory and were worked up here in acid-cleaned glassware. The implication is that two laboratories on separate continents would have acquired the same contaminant, once as mRNA and once as a protein! Micro-sequencing is a legitimate and well-established method and quite powerful in the hands of an expert. Contamination can be a problem, but it is not an insurmountable one, and fear beyond reason must not be allowed to paralyze progress.
>
> We know that our work has been done with care commensurate with the importance of this finding and hope that our result will induce some of our colleagues to look at the enormous reservoir of proteins in our marine invertebrates. Thanks to the editor for allowing us to respond to the readers' concerns.
>
> Christian Schwabe, Danielle Georges.
> FASEB J. Dec 1999, Vol. 13, p. 2338

Now that tunicates as well have a bridgehead on land, it still would not seem reasonable to draw the intermediates between a tunicate and a pig and then revert from a pig to a whale or to construct any common ancestry via this route. Nor is it plausible to say that relaxin in the pig evolved to its present state via evolutionary pressure in the pig when the gene had already been present at the tunicate stage. This was troublesome for the traditionalist. One scientific foray into the world of marine invertebrates provided a fascinating result; what would a systematic survey bring out? It would, I am sure, make an even more persuasive case against evolution via descent with variation. Species and proteins do not support that scheme, and therefore our funding institutions do not support that research.

The Genomic Potential Hypothesis holds species to be immutable, and that contention is strongly supported by protein structures. To wit, all members of a species have the same molecules no matter how long they have lived in different parts of the world and members of different species have a larger proportion of different molecules no matter how long they have shared the same living space. This is a strong prediction, true up to the level of the occasional mutation in a species. There are no molecular fossils that could provide clear answers, but a survivor from 250 million years ago, a recently revived halophile,[16] was not different from contemporary bacteria!

How did the observed pattern of protein distribution come about? Cells are assembled as sets rather than individuals from precellular catalytic units, comprising nuclear coding material with rudimentary protein synthetic capability. These focal points function by the same mechanisms, but take up varying amounts of potential coding material, so that upon cell closure clones with different developmental potential will exist. Regional restriction of coding sequences will by and large give rise to a certain group of phenotypes of one complexity level which will have similar and occa-

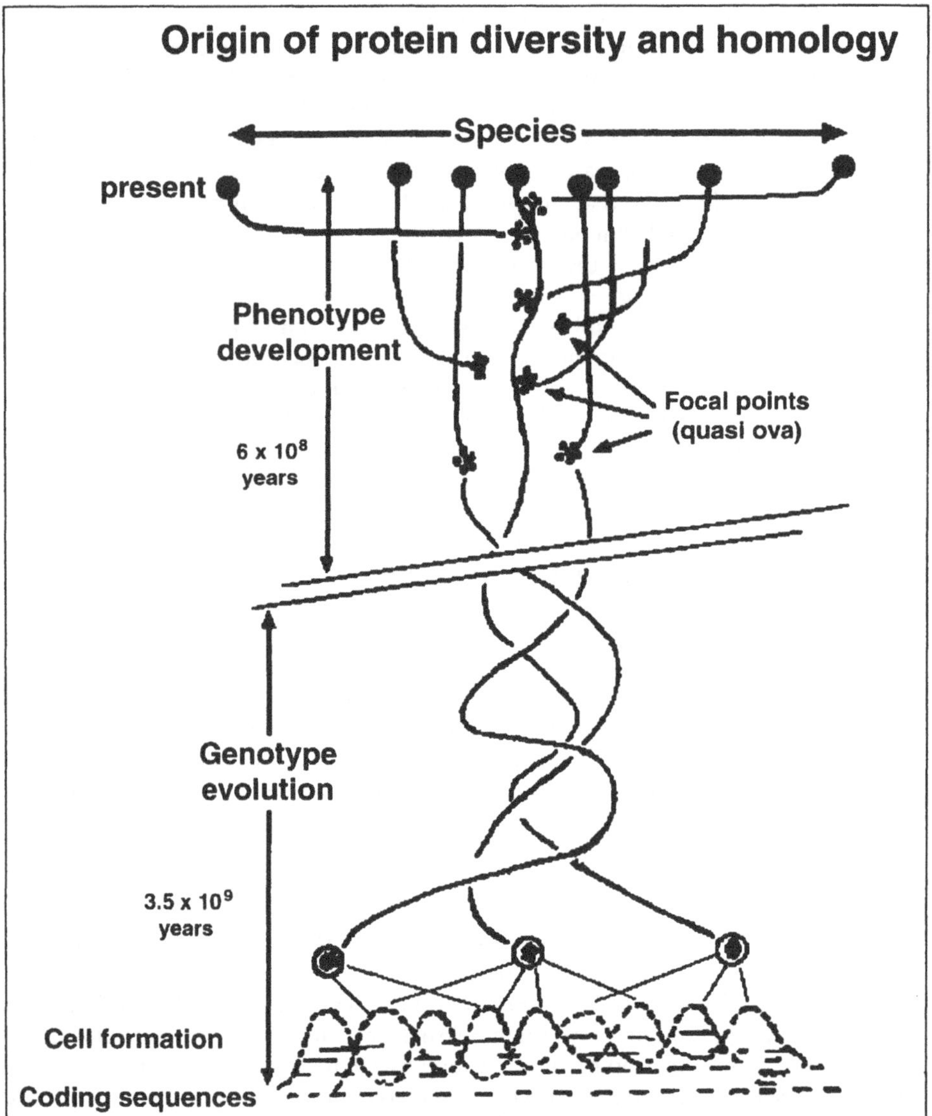


Fig. 13.5. This figure is a summary of the various postulates of the genomic potential hypothesis. Coding sequences evolve from straight chemistry in innumerable pools on the earth's surface to provide a nearly unlimited reservoir of nucleic acid polymers which form themes and variations that are taken up by the various foci of cell formation. Once a cell has formed, it never loses its basic character, remaining either a micro-organism or developing a potential for a large animal during this 3.5 million years of the Archean period which would show up at varying times after the Cambrian period. Each new species eventually appears from its own pro- form and persists until the present or extinction.

sionally identical proteins as noted in the cytochrome table, for example, but do not preclude the "out of line" structures one observes off and on. Each biogenic focus has its own worldline which is contiguous as it transgresses the cellular eon and the post-Cambrian transformation to macro-organism. All biogenic foci, meaning the potential for all species, had self assembled during the same time period, 3.5 billion years ago, so that there is no time segregation in terms of molecular content but rather one of localization. At the bottom of Fig. 13.5 are broken black lines that symbolize oligonucleotides and the bell-shaped ones represent polynucleotide distribution around certain themes. From these polymers cells were formed and the lines leading to the cells are suggesting that cell assembly might involve polymers of several nucleic acid distributions. They all start their organizational march through the Precambrian at the same time and it is only the ripening of phenotypes that staggers their appearance in the fossil record.

Molecular structures do not redraw the Darwinian tree of evolution but rather the clonal distribution of the Genomic Potential Hypothesis, and that forces us to realize that the model of descent with variation cannot be correct. Instead one observes clusters within clusters, both, in taxon development and molecular sequence similarity. One sees parallel world lines of proteins and species evolving as if from nothing, at different levels in antiquity, ending in extinction or breaking the surface to the present.

Someone will call us back to earth by pointing at journals called *Science* or *Nature* and all the evolutionary trees appearing in them that represent the top 5% of the best papers submitted which are replete with evolutionary trees of molecules and creatures. The crème de la crème... could they be wrong? To gain a different perspective one should consider that Agassiz and Owen would have published in those journals! Popularity is not synonymous with truth. Fundamentally new concepts always seem to enter from the outside, starting with a single voice and intense unpopularity; conceptual science is a lonely affair.[17]

Science is naturally divided into experimental and conceptual branches and these intellectual tsunamis surface in about 100-200 year intervals. Experimenters work within the new frames of references until enough has been done to show that a worldview and reality do not match. Within the old model the experimental work reported in these journals is fine, it just does not fit reality which is unperturbed by our formulations. Upon perusal of these papers and a few calculations one gains the distinct impression that branching of species and molecules is impossible; their results actually support the Genomic Potential Hypothesis.

Recently I took a friend through the Smithsonian Museum of Natural History. He read the labels as we approached, one splendid display after another, and then I let him look at them through my virtual magnifiers and, when we left, things were so different, it seemed the venerable Smithsonian had shrugged.

## References


1. Kimura M. A simple method for estimating evolutionary rates of base substitutions through comparative studies of nucleotide sequences. J Mol Evol 1980; 16:111.
2. Cohen J. Will molecular data set the stage for a synthesis? Science 1994; 263:758.
3. Zuckerkandl E. The appearance of new structures and functions in proteins during evolution. J Molec Evol 1975; 7:1.
4. Doolittle RF. The genealogy of some recently evolved vertebrate proteins. TIBS 1985:233.

5. Dayhoff MO. Atlas of protein sequence and structure. The National Biomedical Research Foundation. Vol. 5, 1972.
6. Benton MJ. Diversification and extinction in the history of life. Science 1995; 268:52.
7. Lewin R. Molecular clocks turn a quarter century. Science 1988; 239:561.
8. Wilson AC, Carlson SS, White TJ. Biochemical evolution. Annual review of biochemistry 1977; 46:573.
9. Denton M. Evolution: A Theory in Crisis. Bethesda: Adler & Adler, 1986.
10. Buellesbach EE, Gowan LK, Schwabe C et al. Isolation, purification, and the sequence of relaxin from spiny dogfish (*Squalus acanthias*). Eur J Biochem 1986; 161:335.
11. Buellesbach EE, Schwabe C, Callard IP. Relaxin from an oviparous species, the skate (*Raja erinacea*). Biochem Biophys Res Comm 1987; 143:273.
12. Gowan LK, Reinig JW, Schwabe C et al. On the primary and tertiary structure of relaxin from the sand tiger shark (*Odontaspis taurus*). FEBS Letters 1981; 129:80.
13. Schwabe C, Buellesbach EE. Relaxin and the Fine Structure of Proteins. Austin and Heidelberg: Landes Bioscience/Springer Verlag, 1998.
14. Schwabe C, Buellesbach EE, Hayne H et al. Cetacean relaxin: Isolation and sequence of relaxins from *Balaenoptera acutorostrata* and *Balaenoptera edeni*. J Biol Chem 1989; 264:940.
15. Georges D, Schwabe C. Porcine relaxin, a 500 million-year-old hormone? The tunicate *Ciona intestinalis* has porcine relaxin. FASEB 1999; 13:1269.
16. Vreeland HR, Rosenzweig WD, Powers DW. Isolation of a 250 million-year old halotolerant bacterium from a primary salt crystal. Nature 2000; 407:897.
17. Kuhn T. The Structure of Scientific Revolutions. Chicago: University Press Chicago, 1970.

# Chapter 14

# Experiments in Evolution

Conceptual science interfaces with the experimental world as predictions emanate from the internal logic of a paradigm. Hypotheses concerning the origin and unfolding of life have their first brush with reality when they are held up against fossils. Molecular genealogy is the first prediction evolving from the Darwinian paradigm of "descent with variation", and fossil and protein lineage must confirm each other if the original idea is correct. Evolution by random mutations amounts to an intrusion of divinity into science, and it is satisfying that the concept does not yield acceptable results. The next prediction of the old model pertains to the mechanism of the evolution-induced genomic changes and that is amenable to experimental testing. New technology has made it possible to progress to a deeper level of the genealogy, the rationale for mutation acceptance.

The accumulation of the primordial genome is a chance event by any hypothesis whereby one must grant the possibility that there may have been a tendency to form a certain sequence faster than another, but that problem is left waiting until one knows how primordial condensation occurred. In any case, the genomist claims that all variations observed today are due to prebiotic events[1] in contrast to the old model.[2] Proteins would change continuously in the Darwinian system with survival as the only selective force. Proteins in this type of study come, for obvious reasons, exclusively from survivors. The testable aspect of the hypothesis is the proposal that functionally important amino acids remain constant in a protein and that functionally unimportant ones are subject to mutational replacement.

For easy reference the primary sequences of homologous proteins (relaxins for example) from different species are shown side by side for counting the number of differences.[3] Here significant assumptions enter the picture, namely that all molecules have once been alike and that differences in amino acids in defined positions in a protein are due to mutations.[4] Amino acids that remained unchanged are important for biological functions because their exchange would have more or less severely compromised the owner of such mutants. The primary structures of a series of relaxins shown in Fig. 14.1. have been aligned at the cysteine (C) residues (cross-links) which are identical in all relaxins and insulins except for mouse relaxin.[3]

The letters stand for specific amino acids and the identities of letters in certain positions are scored when we talk about levels of identity between two or more proteins. One then observes which positions (say 12, 13, and 17 in the B chain) have identical amino acids in all relaxin molecules. These are called "preserved" in the old model and that goes along with the assumption that all relaxins come from one ancestor gene and that all uncritical residues have changed due to mutational events.[5]

*The Genomic Potential Hypothesis: A Chemist's View of the Origins, Evolution and Unfolding of Life*, by Christian Schwabe. ©2001 Eurekah.com

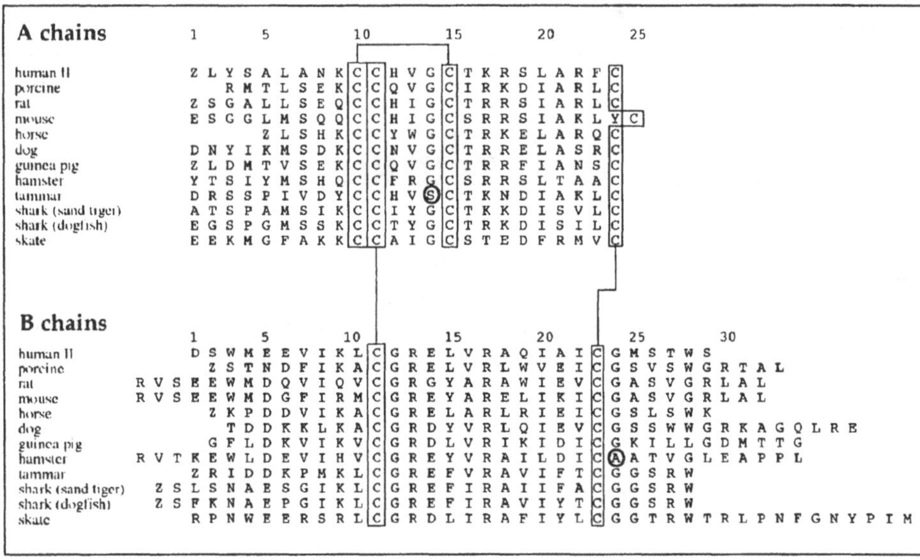


Fig. 14.1. This figure displays the majority of known relaxin molecules. The sequences are aligned at the cross-linking cysteine residues which are constant except for the C-terminal cysteine in the mouse relaxin A-chain. The length of the chains is not of functional significance. The proteins are made in one chain and the loop that connects the C-terminal end of the B-chain (at the right) with the N-terminal end of the A-chain (at the left) is removed by enzymes, and that process depends upon the amino acid sequence of the connecting peptide. The enzymatic process results in a distinct chain-length for relaxins in different species. The active core of these molecules is located between the crosslinks. The receptor binding residues are arginines (R) in positions B13 and B17, and as the eye glides down the columns at these positions one notes R residues only. The various features are discussed in relation to the problems of molecular genealogy.

Furthermore, the neo-Darwinian hypothesis says that the insulin gene had been duplicated long ago and that the "left-over" copy has mutated into a relaxin.[6] Once the astronomical number of mutations had led to an active relaxin any further mutation that would hit the invariant positions would have killed or severely handicapped the owner of that hormone, and therefore all constant residues are important. My colleague Dr. Erika Büllesbach has developed an ingenious as well as practical way to synthesize relaxin and insulin for our NIH-funded research.[3] These derivatives allowed us to experimentally test some of the postulated mechanisms in the Darwinian model of molecular evolution.

Human relaxin was synthesized with either L-alanine or the unnatural D-alanine, substituting for the constant glycines in position 12 in the B chain, and we found that the modified molecules were just as active as the native ones. Furthermore, the constant glycine in B24 could be replaced by alanine with only minor disturbance![3] Yet only once so far has alanine been observed in a natural relaxin (hamster relaxin). From these experiments it follows that the glycines in position B12 and B24 must be constant for reasons other than functionality (Fig. 14.1).

Next, glycine 14 in the A-chain loop (Fig. 14.1) was exchanged for isoleucine (a large molecule) and this change inactivated the hormone.[7] Here then was a case where a constantly appearing residue was also required from a functional point of view.

However probing further, a relaxin with an alanine (slightly larger than glycine) in that position would function very well! Again, only once do we see another fairly small amino acid (serine) in tammar relaxin in that position but alanine, a quite common amino acid in proteins, has not yet been seen in position A 14.

The arginines in the B chain, however, told a different story. They could not be replaced by anything without destroying the hormonal activity of relaxin. The constant arginines in the B chain proved to be the receptor-interacting site, and to remove them and replace them with anything else would be like grinding the serrated edge of a key. The hormone would no longer interact with its receptor.[8] Here then is a case where the constant residues are required for functionality, and our experiments have therefore led to the clear-cut conclusion that functionally important residues are constant but that not all constant residues are functionally important! The clarion is muffled when one learns that humans apparently do not need relaxin so that the relaxin molecule should be under no "pressure" to retain its active structure. The argument that an arginine to X exchange in relaxin would kill (a human) is not tenable! What keeps nonfunctional residues constant and what keeps functional residues constant when the whole molecule is not important for survival? Biology seems to run on rails, the controls that keep us alive keep us from changing, and they do so for important and unimportant amino acids equally well. The experiments favor the Genomic Potential Hypothesis, which places the creation of variety into the prebiotic era and makes biology steady.

Were an inactivating mutation to hit insulin, the victim would die, but by no stretch of the imagination can one extrapolate this observation to relaxin. Furthermore, insulin has at least 35% replaceable residues but differs only very little among mammals (except for the guinea pig). In this case a large number of 'replaceable residues' have remained constant without "evolutionary pressure" to do so, and that also points to stability as the ground cord of life.

Experimenting with whale and pig relaxin, we found the two molecules to be all but identical which was sensational against the background of high variability of relaxins among terrestrial mammals.[9] Because the pig and whale fossils are already distinct in 50 million-year-old layers we looked for a reason why, in spite of numerous changeable positions, their relaxins have not changed during that period with respect to each other. In fact, the relaxins of so-called closely related animals are so dramatically different that the pig/whale similarity is truly astounding. This observation invited another set of experiments which, for convenience, was targeted to the N-terminal A chain end of pig relaxin[10] (Fig. 14.1). After careful chemical dissection of the first seven residues that had remained completely constant between whale and pig, the molecule was inactive. We then synthesized the N-terminal pentapeptide of insulin and coupled it to the inactive truncated relaxin, and noted that the relaxin activity returned. In an even more drastic experiment the truncated relaxin was coupled to a penta-alanine peptide and that too was sufficient to restore the biological activity of relaxin. The functional requirement was merely that a chain be present that could form an $\alpha$-helix. Between whale and pig the relaxin N-terminal ends of the A chain had remained constant for 50 million years, ostensibly without interbreeding, even though this region of the molecule could have mutated to many amino acids without destroying the function of relaxin.

Meanwhile we discovered porcine relaxin in tunicates (Chapter 13) and again one wonders what keeps these amino acids constant against the background of purportedly constant mutational activity which is credited with creating all the variety of life on earth?[11] The gene for "porcine relaxin" was known since the very beginning of the age of animals; it did not need to be invented by mammals, and again, the minimal message is that fossil and molecule genealogy do not match.

Nature is an honest adversary. She never lies but often hedges, and like a good witness, she will never say more than the question warrants. In the case of mouse relaxin she was explicit. When one compares the mouse relaxin in Figure 14.1 with rat relaxin one notices that the C-terminal cysteine in the mouse hormone A chain is displaced outward by one residue.[12] This change makes mouse relaxin different from any other relaxin.

The impact of this finding on the Darwinian picture of the evolution of rodents is quite interesting. The story suggests that physical obstacles once upon a time caused the original primitive rodent ancestor population to split into hares, squirrels, the guinea pigs, and the muridae (mouse and rat). These events are thought to have occurred many million years ago, but the murine split into rat and mouse was a relatively recent event. Of course, one would notice that the segregation scheme is kind of compromised by the fact that the three taxa continued to coexist in the same landscape, but we will ignore that to keep the story flowing. Guinea pig relaxin was synthesized in our laboratory and found it to be very active in that species.[3] We synthesized rat relaxin as well, which proved to be one of the very active relaxins, and both of these rodent molecules have the regular disulfide bond structure. Since both, rat and guinea pig, are purportedly ancestral, a normal disulfide bond pattern is the older one and the mouse, after separating from the rat, must have converted a good relaxin gene to one with an extra residue in the A chain (according to neo-Darwinists). We have synthesized the regular mouse relaxin and noticed relatively low bioactivity in the mouse. Curiosity caused us to synthesize the mouse relaxin minus the tyrosine (Y) (Fig. 14.1) that had purportedly been inserted after the mouse/rat separation.[12] Biological activity measurements in mouse tissue showed clearly that the synthetically "reverted" mouse relaxin is superior to the real mouse relaxin. According to the old paradigm the mouse gene had suffered an insertion mutation and the inferior relaxin had propagated against selection pressure through the whole mouse taxon to the exclusion of the better gene. Darwinian selection would, by definition, always drift toward the better molecule which is how evolution to complexity is envisioned.[13] Here nature tells us, with rare frankness, that idea is wrong! Genomists agree, the mouse never had a different gene, and that gene was sufficient to provide us with all the mice we need.

Neither proteins nor the encoding genome have sensors for needs or directionality. In contrast, life has developed an unmatched system of control and repairs which keeps the genome free of mismatches. That status quo-protecting system is not judgmental as concerns the quality of a gene. Conversely, the genome has no means of eliminating weakly-active molecules or to improve them, but because of the natural redundancy of functions in all living systems loss of a particular function on account of mutational activity can, in some cases, be overcome. This has become particularly clear with the advent of gene knock-out technology which has enabled us to remove an "important" gene only to find that in some cases the animal survives happily via compensatory functions. A whole gene has been kept constant without a need for it in the organism.

All those basic residues in the relaxin A chains (R and K in Fig. 14.1), the high isoelectric point (high net positive charge), was all of that necessary? Human relaxin was synthesized with all four basic residues in the A chain, replaced by the neutral unnatural amino acid citrulline. The modification lowered the isoelectric point into the acidic range but had no effect on the relaxin activity! The relaxin receptor recognized this molecule as well as an unmodified relaxin, and the idea of evolutionary pressure causing all relaxins to retain global molecular features, such as the isoelectric point, seemed seriously unconvincing.

Almost all cells in an organism have insulin receptors as well as receptors for many other hormones, but relatively few have relaxin receptors. The process whereby the specific receptor distribution came about is far beyond our present understanding of biology. It has been suggested in the literature that the insulin gene long ago duplicated to give rise to a relaxin gene via a stream of mutations.[6] A new function , however, is thought to be targeted, i.e., the receptor should be present and the hormone should mutate to match this target (or vice versa).[14] That means that one of the members of the receptor/hormone pair mutates to a new function without a target.

Not really, they say, because at the same time and in the same cells, the insulin receptor gene duplicates and when only this duplicate picks up mutations to produce a relaxin receptor, we have our target. Meanwhile the old insulin receptor stays intact because all cells need carbohydrate metabolism, for example. Clearly, this "conjugated miracle" model of targeted development cries for divine intervention, and it is impossible to point out how many noncontiguous and anticipatory changes had to occur to bring these events to a proper conclusion. If one would want to back away from divinity and argue chance, such as in mutations, one would need to consider the basic reality of this process. For the 50 some amino acids of insulin it would take $20 \times 10^{50}$ trials to explore all random possibilities on the way to a relaxin structure. Receptors contain more than 1000 amino acids so that the conversion of the insulin (or any other) receptor to a relaxin receptor would require about $20 \times 10^{1000}$ trials. This would mean about $4 \times 10^{1050}$ trials for the development of this receptor hormone pair. Probabilities such as 1 in $10^{1000}$ simply mean that such an event would not happen. The number of failures would fill the universe many times.

What kind of phenomenon could keep nonessential residues constant over millions of years if change is the stew that nourishes novelty?[15] The Genomic Potential Hypothesis says that there is no stew, stability is the standard and changes are accidents. How is it possible that many proteins in different species look alike to some extent without functional needs to keep them so? The new view is that functions are recruited from untargeted pools of similar nucleic acid sequences that can appear in many organisms and that do not change during development. Some proteins taken from bacteria or protozoa are active in the mammalian cell, and even a transcription factor from archaebacteria interacts properly with human DNA.[16, 17] Some proteins have a mosaic pattern of runs of 20-30 amino acids identical to homologous proteins in other species, and other stretches of amino acids that are matching several stretches of proteins with totally different functions.[18, 19] These observations, I think, are a telltale sign of a generic origin of life, a clonal affair based upon nucleic acid chemistry that coalesced in many places into biogenic droplets like fog condenses on the ground.

All observations such as gene duplications and redundancies are primordial chemical events that have been converted to mutations, gene duplications, and lateral gene transfers by the Darwinian/neo-Darwinian ideology.

These experiments had been performed in my laboratory but the literature is full of such examples. Why do others not see the same phenomenon and acknowledge the problem? Perhaps they were primed to look at their results from the platform of the old prejudices. Frank Plumpton Ramsey[20] and his theory tells us in brief that any large enough, apparently random, collection of items will contain an orderly substructure, and that the complexity of possible substructures depends upon the number of members of a basic set.

Lately the age of the genetic code was researched[21] using a statistical analysis of tRNA sequence relatedness. Data in the literature plus assumptions about the inverse of the mutation accumulation rate (sort of a biological Hubble constant[22]) were used to deduce a focal point where all tRNA sequences would become one and the same, the point where the genetic code was created. The time at which this happened clearly depended upon the meaning given to the sequence differences between the contemporary tRNAs, which in turn depended upon the paradigm that had produced the differences via mutations.

The uniformity of the genetic code in all living creatures examined so far has always been interpreted as evidence for a single-point origin.[21] The paper offers the conclusion that the genetic code might be older than life itself, but not so old that one would need to presume an extraterrestrial origin. The conclusion supports the Genomic Potential Hypothesis (Chapter 6), but it does not follow from the paper unless the Darwinian basis is taken as self-evident.

The author of the *Science* paper admitted, in an answer to my argument, that an original distribution of tRNAs, just as they are seen today, would invalidate his conclusions, but he was sure that "God is not malicious".

The sequence differences between the many tRNAs that these investigators examined are a fact. The evolutionary distances between the species, i.e., the phylogeny of eubacteria, archaebacteria, and eukaryotes, are based on the idea that they were derived from each other by mutations. This means that the tRNA distances, indicated by the sequence differences that are supposedly an indicator of branching, are in fact a restatement of a hypothesis. Thus, we have no independent means other than a paradigm to tie together the various tRNAs into a common ancestor which most likely never existed. This intrusion of reality made no impression upon the investigators who assert that: "kinship relations are revealed by alignment of sequences", whereas only similarities are revealed by such a comparison. Kinship is a derived property that comes from a paradigm; there are no independent means whereby one can determine how these tRNAs came about, there are no fossil molecules.

These studies, like those mentioned in the previous chapter, have appeared in a very prestigious journal where many of this type of papers are published because editors are under palpable "paradigm pressure" and are often uncomfortable in matters of epistemology. What makes investigators accept a model so uncritically and, more importantly yet, what makes them force fit data into such a model? Here, I think, a biological uncertainty principle comes into play, which states that our mind cannot view a set of data without at once seeking for an underlying order. More often than not the order is achieved according to an internal (intuitive) set of parameters, and when mental coziness spreads, stars and molecules will move into the proper positions to satisfy our pictures. Ramsey's theory explains this phenomenon.[20]

The reader has witnessed a head-on collision of experimental evidence with a major postulate of the neo-Darwinian model. As a consequence, differences are no longer mutations but rather have reverted to just differences. Both, gene duplications and mutations, are mimicked by primordial variability that was stabilized by the constraints of biology.

## References


1. Schwabe C. Evolution and chaos. Computers Math Applic 1990; 20:287.
2. Benton MJ. Diversification and extinction in the history of life. Science 1995; 268:52.
3. Schwabe C, Buellesbach EE. Relaxin. Heidelberg: Springer Verlag, 1998.
4. Mayr E. Darwin's influence on modern thought. Sci Am 2000:79.
5. Samols D, Thornton CG, Murtif VL et al. Evolutionary conservation among biotin enzymes. J Biol Chem 1988; 263:6461.
6. Dayhoff MO. Atlas of protein sequence and structure. The National Biomedical Research Foundation. Vol.5, 1972.
7. Buellesbach EE, Schwabe C. Structural contribution of the A chain loop in relaxin. Int J Peptide Protein Res 1995; 46:238.
8. Buellesbach EE, Yang S, Schwabe C. The receptor-binding site of human relaxin II. J Biol Chem 1992; 267:22957.
9. Schwabe C, Buellesbach EE, Hayne H et al. Cetacean relaxin: Isolation and sequence of relaxins from *Balaenoptera acutorostrata* and *Balaenoptera edeni*. J Biol Chem 1989; 264:940.
10. Buellesbach EE, Schwabe C. Relaxin structure: N-terminal A-chain helix. J Biol Chem 1987; 262:12496.
11. Georges D, Schwabe C. Porcine relaxin, a 500 million-year-old hormone? The tunicate *Ciona intestinalis* has porcine relaxin. FASEB 1999; 13:1269.
12. Buellesbach EE, Schwabe C. Mouse relaxin: Synthesis and biological activity of the first relaxin with an unusual crosslinking pattern. Biochem Biophys Res Commu 1993; 196:311.
13. Mayr E. Evolution. Sci Am 1978; 239:45.
14. Shapiro JA. Adaptive mutation: Who's really in the garden? Science 1995; 268:373.
15. Ohno S. Dispensable genes. Trends in Genetics 1985; 1:160.
16. Garrett RA, Dalgaard J, Larsen N et al. Archaeal rRNA operons. TIBS 1991; 16:22.
17. Pennisi E. Static evolution. Sci News 1994; 145:168.
18. Baron M, Norman DG, Campbell ID. Protein modules. TIBS 1991; 16:13.
19. Yonekura H, Nata K, Eatanabe T et al. Mosaic evolution of prepropancreatic polypeptide. J Biol Chem 1988; 263:2990.
20. Graham RL, Spencer JH. Ramsey's theory. Sci Am 1990; 263:112.
21. Eigen M, Lindemann BF, Tietze M et al. How old is the genetic code? Statistical geometry of tRNA provides the answer. Science 1989; 244:673.
22. Osterbrock DE, Gwinn JA, Brashear RS. Edwin Hubble and the expanding universe. Sci Am 1993:84.

# Chapter 15

# Quintessence

Steven Weinberg once said that the complexity of physics today reaches very close to the edge of the human intellect.[1] Where would that leave us, the biochemists and biologists, who study a subject many times more complex than physics? Biology has invented physics, and with that invention it has been possible to chip one equation of physics after another from the "ephemeral eternity" that marks our universe at each instant. Presently physics has reached a level of power and sophistication that makes it possible to predict, with fair confidence, when the universe began, how it started, how the galaxy formed that is the home for our sun and its planets, and where all of it is flying at what speed. We even know at what time our planet will be burned up in the solar corona. From the slowdown of binary systems we can calculate the power of gravity to the 12th decimal place,[2] and we are beginning to uncover 'quintessence', a different force that takes over when gravity diminishes in the distant power fields of space.[3] Is quintessence the gravitational pull of a neighboring universe? You see our protein-based computer is tempted to spark off at the slightest cue and will work on a problem until a new concept can be tacked onto our quilt of science.

Yes, we also know what we are made of. We know the electric fields of atoms that caused molecules to assemble, the rules whereby they assembled to form self-perpetuating units, some of which have reached such complexity as to be able to do all the incredible feats described in the previous paragraph. Should we be satisfied with the suggestion that all of this is based upon chance processes such as mutations? An unending string of lucky mutations at the beginning of quantum theory and relativity is certainly not one of our better ideas.

A hypothesis of evolution exerts influence on the development of biological sciences of which biochemistry in turn provides major parameters for anchoring concepts, connecting them to reality, as it were. During the development of the Darwinian model, the discipline of biochemistry did not exist and genetics had just been started but logic, the very basis of scientific discourse, was there. Thus Darwin replaced God by logic but put chance, the hand of divinity, right back into the core of the model. The Genomic Potential Hypothesis purges constructive chance events from the equation and returns the problem to biochemistry in the true sense of the term. As a consequence evolution seems more complex but, in exchange, it has become a legitimate target of scientific pursuit. Instead of tranquilizing the human mind with fortunate accidents, the new direction calls for answers. Is there a higher order code in the genome, did the beginning nucleic acid polymer show some sequence-dependent tendency to collect compatible sequences to make possible that one full percent of nucleic material of the human genome would carry functioning messages? How is morphology expressed, how do sets of dependent functions get organized such as food intake, digestion distribution, or reproduction or locomotion? These functions are all finished when the first macro-organisms enter the fossil beds; they were assembled by rules other than needs that were still unexpressed. The complexity of these interactions is staggering but they came about by self-assembly based upon structural features, i.e., they are ultimately discoverable! Questions of this kind will guide the research effort of evolutionists back to the problems that are subject to the falsification test.

The phrase 'in principle' falsifiable[4] or discoverable means that a model is contiguous. Interaction of historically uncoupled events (chance) are by definition unpredictable regardless of whatever level of supernatural expertise one might bring to bear on such a problem. For example, the old theory holds that the organisms present at the Cambrian edge represent the total biological pool and that among them must be the stock for the whole post-Cambrian world, for the fauna and flora that developed from these early life forms by gene duplication and mutations.[5] Inspecting a genome of one of these animals closely one would not be able to predict where or if the required mutations would occur to produce a new species. And that is not a matter of ignorance or overwhelming complexity but rather of principle. In other words, even if we knew a genome to the most profound level and with supernormal facilities, nothing in that structure can tell us what mutation would hit next. In this case predictions become "in principle impossible" and to make respectable extrapolations from such a process is, in fact, impossible. Causality and contiguity are the quintessence of science and mutations are not contiguous. They cannot be constructive for all the reasons given in this book, and it does not matter that everybody in evolution seems quite happy with mutations as the modulator of biological form. Science is, in the final analysis, decided neither by majorities nor by prominence of the proponent. Not too long ago a mathematics-supported origin of life scenario was published that, at the end, decayed into a mutation-driven diversification scheme.[6]

It takes but one mutation to cause a disease and that mutation does not even need to occur at one specific place in a gene. About 30 different mutations have been identified that cause osteogenesis imperfecta, and the same is true for many other genetic diseases. If it were possible to convert a zebra into a horse (to keep it impersonal), how many mutations would it take and how precisely would they have to be placed and in what sequence must they occur and lastly, how many deadly mutations must the convert avoid? So, why is it not reasonable to build life from a series of lucky reactions and sort out that which failed? Because the number of failures would be larger than the number of particles in this universe!

Perhaps the gambler can make a positive contribution to this discussion. Dice have only 6 surfaces and the chance of winning is 1:6, which by biological standards are high odds. The odds of losing, of course, are 5 times higher so his fate, which cannot be influenced by skill, is no surprise. Although frowned upon in that trade, "fate" can be influenced by breaking the symmetry of the system (loading the dice). Nature has done it legitimately and the Genomic Potential Hypothesis is based upon the realization that the 'dice' have been loaded when energy condensed into atomic structures. In such a setting randomness comes about only when there is no structural discriminator in the interacting system.

The loaded dice will not show the winning number at every throw but will do so rather more frequently than an unmodified one. That is precisely what happens in chemistry where complex reactions always yield an approximately normal distribution around a major product and where repeats under similar conditions are a very likely event, hence the multiple origin scenario.

The way life produced itself is the way it persists. Mass action, affinity constants, and competition produce all our pathways of control in conjunction with stochastic movements of proteins.[7] Affinity constants are the loaded dice in biological control processes and even these can be modulated stochastically by increasing or decreasing the affinity for a target by, say, phosphorylation. The regulatory site of DNA will be occupied by inhibitors or promoters depending upon the amounts and affinities of the various molecules in the nucleus. Of course, this condition holds for other reactions in the cell and explains why life, why the cell needs membranes which prevent loss of solutes and permit control over the concentration of regulatory components. Affinity constants dictate how many molecules have to be in a fixed volume to cause 50% occupancy of the binding site. Mass action and molecular structure have produced organisms in the Genomic Potential Hypothesis and that is the principle whereby they have to function. In a roundabout way cellular metabolism provides a hint as to how self-assembly produced cells during primordial times.

The genomist will look at the first animals and predict that all of them will remain just what they were in the Cambrian until the present or extinction.[8, 9] New forms will come from species-specific precursors at later times. The fossil record gives a nod to the new world.

The Genomic Potential Hypothesis raises the spectre that all species have shown up by now. Any "new" form that is reported off and on has merely escaped notice or is the result of hybridization. Truly new species should be more complex than *H. sapiens (H. sapiens super sapiens)* which we, after deep self-analysis, readily admit to be unlikely. Perhaps potentially more advanced clone members are living among us, analogous to the cohabitation of *H. erectus* and *H. sapiens* on the island of Java. This scenario is not impossible but difficult to confirm considering the extensive global mixing of genes that comes with technology. The Genomic Potential Hypothesis carries within its conceptual core the prediction that evolutionary development is finite. Species wax and wane in numbers but diversity is on a steady decline.

After a 150-year grace period evolutionary biology must get back to basics or risk being ostracized from science by the science it created. We are not playing in a sandbox by ourselves; we cannot forever nurture and protect a miniature pontifical college that will preach to us on matters of evolution and suppress the turmoil, which is such a fertile ground for advancement of knowledge.

## References


1. Weinberg S. The First Three Minutes. New York: Basic Books Inc., 1977.
2. Taylor JH, Fowler LA, McCulloch PM. Measurements of general relativistic effects in the binary pulsar PSR1913+16. Nature 1979; 277:437.
3. Ostriker JP, Steinhardt PJ. Quintessential Universe. Sci Am 2001; 284:47.
4. Miller DW. Popper Selections. Princeton: Princeton University Press, 1985.
5. Benton MJ. Diversification and extinction in the history of life. Science 1995; 268:52.
6. Eigen M, Schuster P. The Hypercycle. New York: Springer Verlag, 1997.
7. Misteli T. Protein dynamics: Implications for nuclear architecture and gene expression. Science 2001; 291:843.
8. Schwabe C, Warr GW. A polphyletic view of evolution: The genetic potential hypothesis. Perspec Bio Med 1984; 27:465.
9. Schwabe C. Evolution and chaos. Computers Math Applic 1990; 20:287.

# Appendix I

# New Problems, New Names

An extensive nomenclature has developed from within the Darwinian paradigm. It will take a while to get weaned from that habit, particularly since the new model, by its nature, is austere in that regard. In as much as the genomic material, as the basic constituent of life, is the center of the new model, **genomic potentialism** or **genomism** may be a shortcut to Genomic Potential Hypothesis, and those who are accepting the basic premise would be **Genomists.**

Differences in the sequence of homologous proteins in different species are derived from the **primordial pools** and are a product of the redundancy of nucleic acid polymerization and reiterations by complementary reproduction of nucleic acid strands. Sequence repeats in different proteins that could not have come about by chance (here I agree with Darwinians) do arise quite naturally from untargeted **reiteration** in the primordial pool.

The new expression for sequence similarity is **primordial repeats.** These may occur in certain segments of the protein or as a mosaic pattern of 20 or so residues in various parts of the chain.

Proteins of different functions, which have sequence similarities such as the insulin-like growth-factor family and the relaxin family of hormones, are **primordial reiteration sequences.** They are primordial nucleic acid strand duplications which would have been sufficiently imprecise (without the biological controls) to produce structural themes and variations. The difference is important because biological gene duplication leads to identical products that would have to be converted to a new sequence by targeted mutations, which are disallowed by epistemology. Biological gene duplication is not strictly impossible and would lead to two or more identical genes. If some of these are different in a few places it is impossible to ascribe them to either the background of random mutations or to primordial events. If two proteins differ enough to have different functions they are certainly **primordial repeats.**

Interspecies differences that involve one or more positions in a homologous protein (formerly mutations) are now **clonal variants.** This term implies that, as one examines available sequence data on many proteins from various species, one could expect to see a pattern of similarity that may suggest **clonal proximity.** The table of cytochrome sequences, in fact, suggests such a trend and if other proteins of the same species show the same degree of similarity then one may speak of **primordial proximity** during biogenesis, and here the order of primates may be an example. We do not have enough structural information in order to obtain a more persuasive picture, but the new age of proteomics may, as a purely academic by-product, provide more information concerning the mode of assembly of species. The enormous reservoir of

*The Genomic Potential Hypothesis: A Chemist's View of the Origins, Evolution and Unfolding of Life*, by Christian Schwabe. ©2001 Eurekah.com

invertebrate proteins is likely to make a significant contribution to the question concerning the potential of primordial pools. How large a group of life forms can each pool produce? Are large divisions, such as marsupials versus placentals, reflected in our protein record? The result of such studies may tell us about **clonal relatedness.** As we cut loose from the old model and indulge in potentially beneficial uncertainty, a lot of new experimental avenues will open up.

# APPENDIX II

***Synopsis of the New and the Old Model of Evolution***

| The (clonal) Genomic Potential Hypothesis | The (linear) Darwinian Hypothesis |
|---|---|
| Origins of life were an inevitable chemical phenomenon and therefore a massive event. | The origin of life was a chance phenomenon and therefore a singular event. |
| Life aggregated in many, relatively small 'breeder-reactors' which collectively produced the developmental potential for every form of life. | Only one form survived to seed the earth. |
| Life is polyphyletic from its inception, there is no major branching, only survival or extinction. New forms come from newly metamorphosed clones. | Life is monophyletic and continuous branching caused by chance events creates new forms. |
| All self-assembly functions are uniform because there is only one way to form the core of living systems. | Uniformity of certain functions is due to a singular origin. |
| Peripheral variety is the product of primordial repetitive and reiterative polymerization of nucleic acids, i.e., a <u>contiguous process</u> | Variety is the product of biological phenomena, i.e., gene duplication, mutation and selection, and reproductive isolation, i.e. <u>a non-contiguous process</u>. |
| The genomic code is uniform in all clones of primary life forms because it was "structure-induced", i.e., it was the only possible code. | The genomic code is uniform because all organisms come from one origin and the code functions do not mutate as all others functions must do to evolve. |
| Differences between species are a function of the genomic potential of each origin of life. | Differences between species are a function of mutation acceptance |
| Evolution of genotypes occurred during the pre-Cambrian period. | Not much happened until about 600 million years ago. |
| Organisms appear in the fossil record in the sequence in which they finished their genomic evolution, i.e., the simplest first. Many different clones of similar levels of complexity finish simultaneously in many places. | Simple organisms appear first in one locality and spread from there slowly increasing in number and variety. All appear in succession, one from another. |

| The (clonal) Genomic Potential Hypothesis | The (linear) Darwinian Hypothesis |
|---|---|
| The evolutionary tree is an illusion created by the successive metamorphosis of clones followed by the rapid development of clone-specific phenotypes. | The evolutionary tree shows sequential branching of species and proteins as a consequence of chance processes (mutations). |
| There are no intermediate forms. | Every life form is an intermediate between a predecessor and a successor. |
| The world-line of each clone, extinct or extant, originated 3.5 billion years ago approximately. | The world-line of each macro-organism begins after branching and segregation. |
| The evolution of macroscopic forms is due to differing genomic configurations and independent of protein structures. | Different macroscopic forms are due to different genes and proteins. |
| Life anywhere in this universe will be identical to the life forms we are familiar with, i.e., carbon-, nitrogen-, hydrogen-, oxygen- based and will use the same genetic code. | Life on other planets may be completely different in as much as chance processes do not repeat. |

# Index

F



G



H



I



K



L



M

**U**



**V**



**W**



**Z**

For Product Safety Concerns and Information please contact our EU representative GPSR@taylorandfrancis.com Taylor & Francis Verlag GmbH, Kaufingerstraße 24, 80331 München, Germany